Starting a Small Food Processing Enterprise

Peter Fellows, Ernesto Franco and Walter Rios

Practical Action Publishing Ltd
27a Albert Street, Rugby, CV21 2SG, Warwickshire, UK
www.practicalactionpublishing.org

First published in 1996
Reprinted 2003

ISBN 10: 1 85339 323 1
ISBN 13: 978 1 85339 323 5
ISBN Library Ebook: 9781780446028
Book DOI: http://dx.doi.org/10.3362/9781780446028

A catalogue record for this book is available from the British Library.

Since 1974, Practical Action Publishing has published and disseminated books and information in support of international development work throughout the world. Practical Action Publishing is a trading name of Practical Action Publishing Ltd (Company Reg. No. 1159018), the wholly owned publishing company of Practical Action. Practical Action Publishing trades only in support of its parent charity objectives and any profits are covenanted back to Practical Action (Charity Reg. No. 247257, Group VAT Registration No. 880 9924 76).

Typeset by My Word!, Rugby

Technical Centre for Agricultural and Rural Cooperation (ACP-EU)
The Technical Centre for Agricultural and Rural Cooperation (CTA) was established in 1983 under the Lome Convention between the ACP (African, Caribbean and Pacific) Group of States and the European Union Member States. Since 2000 it has operated within the framework of the ACP-EC Cotonou Agreement. CTA's tasks are to develop and provide services that improve access to information for agricultural and rural development, and to strengthen the capacity of ACP countries to produce, acquire, exchange and utilise information in this area. CTA's programmes are organised around four principal themes: developing information management and partnership strategies needed for policy formulation and implementation; promoting contact and exchange of experience; providing ACP partners with information on demand; and strengthening their information and communication capacities.

CTA, Postbus 380, 6700 AJ Wageningen, The Netherlands

Contents

Acknowledgements

The authors would like to express sincere thanks to the many people and organizations who have assisted in the preparation of this booklet. In particular, thanks to Wendy Bullar for the artistic interpretation, and Matthew Whitton for the illustrations. We would also like to thank the following people for their efforts in producing the business plans and for their permission to use them in this publication:

Bertha Msora, Ranche House College, Harare, Zimbabwe

Diana Colquichagua, ITDG, Lima, Peru

Anita Manandhar, Centre for Rural Technology, Kathmandu, Nepal

Finally our grateful acknowledgement of financial support from CTA – the Technical Centre for Agricultural and Rural Cooperation ACP-EU – and the Overseas Development Administration of the British Government.

Peter Fellows
Ernesto Franco
Walter Rios

About the authors

Peter Fellows is a consultant and can be contacted at Midway, 14 Farndon Road, Woodford Halse, Northants, UK.

Ernesto Franco and Walter Rios both work for IT Peru; contact them c/o Intermediate Technology, Myson House, Railway Terrace, Rugby CV21 3HT, UK.

Technical Centre for Agricultural and Rural Cooperation (ACP-EU)

The ACP-EU Technical Centre for Agricultural and Rural Cooperation (CTA) was established in 1983 under the Lomé Convention between the African, Caribbean and Pacific (ACP) States and the European Union Member States.

CTA's tasks are to develop and provide services that improve access to information for agricultural and rural development, and to strengthen the capacity of ACP countries to produce, acquire, exchange and utilize information in these areas. CTA's programmes are organized around three principal themes: strengthening ACP information capabilities, promoting contact and exchange of information among partner organizations and providing information on demand.

CTA, Postbus 380, 6700 AJ Wageningen, The Netherlands

The purpose of this book

Small-scale food processing offers good opportunities for enterprising people to generate income and employment using locally available resources. Raw materials are often readily available, the level of technology and skill needed to operate a process is usually obtainable, and processed food products often have a good market.

However, none of these factors should be taken for granted when thinking about starting a small business. There are many routes to failure. In particular, it is not sufficient to know how to produce a high quality food: the producer must also know how to sell it effectively and how to control the financial side of the business. For long-term sustainability, the producer must also know how to plan and develop the business.

This book brings together important aspects of both the technological and business skills needed to start and operate a small food processing business successfully. The emphasis is on thorough planning before the enterprise is established and then on careful control of production to minimize costs and maintain the desired product quality.

There are different definitions of micro- and small-scale enterprises, but in this book they are considered as follows:

- A micro-enterprise is one in which the owners themselves work each day to produce foods in the enterprise and also manage it. The number of workers does not exceed 10 and the annual sales are less than $12,000.
- A small-scale enterprise has an owner or manager who may not work at the production site. The number of workers is under 20 and the annual sales are less than $25,000.

This book is intended as a training aid for extension workers and small enterprise development organizations, although others involved in rural and urban development programmes will also find the information useful.

The text is arranged in six chapters, each describing an important aspect of the operation of a small food business. Each of the main topics described is interpreted using an illustration accompanied by a caption. The illustrations may be used by extension workers and trainers to form posters and the captions can be translated into local languages for greater effectiveness in reaching small-scale producers. Examples of business plans are reproduced in Appendix 3, together with comments from bankers and consultants who have reviewed the plans.

This book complements other publications by IT, including:

Making Safe Food – a book describing good hygienic practices and safe food handling (40 pages – contact IT, Rugby).

Small-scale Food Processing: A guide to appropriate equipment – a catalogue of small food processing equipment from around the world (160 pages – contact IT Publications, London).

ALMOST ANYONE can be an entrepreneur, but some people are better at operating a business than others. A checklist of entrepreneurial characteristics is included in Appendix 1.

Different levels of entrepreneurial skills are needed, depending on the level of technology and the environment in which a business operates. The diagram below* indicates that zone A provides the easiest conditions while zone D requires the greatest entrepreneurial skills. It is important to note, however, that all businesses involve taking risks and potential entrepreneurs should be made aware of these risks as well as the likely rewards.

SIMPLE TECHNOLOGY

A

- steady market
- low competition
- long contracts
- assured supplies
- technology simple

B

- competitor or customer behaviour uncertain
- fluctuating orders
- supplies not certain
- technology simple but changing rapidly

STABLE BUSINESS ENVIRONMENT ← → **UNCERTAIN BUSINESS ENVIRONMENT**

C

- competition not great
- orders from a range of customers and no monthly fluctuations
- complex but predictable technology

D

- large number of active competitors
- supplies from many sources
- complex product and process technology
- high rate of technological change

COMPLEX TECHNOLOGY

* with acknowledgement to Garry Whitby

ENTREPRENEURS usually have some basic ideas of the foods that they can produce according to their own experience of where they live. Often the idea comes from seeing others successfully producing a food and then trying to copy them. When someone has an idea for a small business he/she will often try to start up straight away without having thought clearly about the different aspects involved in actually running the business. This 'hit and miss' approach often results in failure during the first or second year of the business.

To reduce this risk of failure it is necessary first to decide whether the idea is feasible (is it likely that the small business will be successful?). This involves doing a short **market survey** and it will usually involve producing a **feasibility study** and possibly a **business plan** (Chapter 4). A feasibility study is a good way of working out the likely success of a business on paper before scarce resources are spent in actually getting started.

Use a market survey and feasibility study to decide which products to make

Start with the customer

Assuming that the entrepreneur has an idea for a product, the first thing to do is to find out from potential customers what level of demand there is for the food (or what is 'the market' for the food). A simple market survey should be carried out by talking to people who are expected to buy the product. Ask them how much they will buy, how often and for what price.

The market study should be a short exercise so as to keep the costs low. In-depth market research is not necessary in most situations.

From the information gathered from potential customers the entrepreneur can work out the total demand for the product. This involves asking a number of questions such as:

1. Are the people interviewed really representative of all potential customers?
2. How many potential customers are there in total?
3. Will people in different income groups buy different amounts of food or at different frequencies?

Table 1 (page 8) shows the results of such a survey from Africa.

Find out about the demand for your product by asking potential customers

Table 1: Survey of potential demand for cooking oil in a rural town
(a) Size of the market

Type of customer	*Number in each category**	*Amount of oil bought per customer (litres per month)***	*Total demand (litres per month)*
Low income	22,400	0.4	8,960
Medium income	1,768	2.5	4,420
High income	260	3.6	936
TOTAL			14,316 litres

* from official statistics for the town
** average of information given by 50 customers interviewed

(b) Value of the market

Type of customer	*Amount of oil bought each purchase (litres)*	*Cost per litre of oil ($)*	*Number of litres per month*	*Value of market ($ per month)*
Low income	0.1 (100 ml)	5*	8,960	44,800
Medium income	1.0	4	4,420	17,680
High income	1.0	4	936	3,744
TOTAL				$66,224

**Oil is more expensive when bought in small quantities.*

It might be helpful to ask potential customers about where they buy the product now, what packaging they prefer, what is good or bad about the quality of the product and what they would like to change about it. The entrepreneur also needs to think about how the product is actually going to be sold and what the competitors are doing. This is the next stage in determining the market.

Different ways of selling your product:
(a) customers come to the factory
(b) through shops or supermarkets
(c) in the market place
(d) at a market stall

Competitors

Competitors are very important to the success or failure of a new business and the entrepreneur should know:

- Who is producing similar products?
- Where are the competitors?
- What is the quality and price of their products?
- What offers or incentives do they give to retailers?
- What can the entrepreneur do to make the new product better?
- Why would a customer change to buy the new product?
- What are competitors likely to do if a new product is introduced?

The entrepreneur then needs to decide how much of the estimated demand can be met by their production. Table 2 gives a general indication of the percentage of total demand that can be met by new businesses, depending on the number and size of competitors.

In summary, the entrepreneur can use all this information to decide which food to make, how much food he/she will have to make each week to meet a known share of the total demand and where there is a chance of competing against other producers. The main questions are: 'Is it a good idea?' and 'Shall I go ahead and invest in the business?'

Table 2: Estimates of market share for a new business with different levels of competition

Number of other producers	*Many*				*Few*				*One*				*None*
Size of competitors	*L*		*Sm*		*L*		*Sm*		*L*		*Sm*		
Product range	*S*	*D*	*S*	*D*	*S*	*D*	*S*	*D*	*S*	*D*	*S*	*D*	
Market share (%)	0-2.5	0-5	5-10	10-15	0-2.5	5-10	10-15	20-30	0-5	10-15	30-50	40-80	100

Note: L = Large, Sm = small, S = similar, D = dissimilar
Originally published in *Do your own scheme: a manual for the entrepreneur* by Small Business Promotion Project, Nepal Ministry of Industry, and GTZ.

Competitors are a vital factor in the success or failure of a business

A note about new products

There are obvious advantages in making a product that is new to an area (for example, there will be no competitors to begin with). However, the entrepreneur should be very sure that people will buy the new food at the expected price. This needs a trial production to make samples for test-marketing (for potential customers to taste the food and give their opinion). All of this takes longer and costs more than making products that are already known. In addition, up to 80 per cent of new products fail within the first year, so the risks are higher and obtaining a loan may be more difficult.

A good compromise is often found by modifying an existing product to create something different which would appeal to a new market.

Not all new products are successful; market research is needed to make sure that a demand exists

Good quality foods and good presentation are essential to get good sales

IF THE BUSINESS IDEA seems promising, the entrepreneur then has to find out if he/she is capable of producing the food in the required amounts and at the correct quality and price. It is also necessary to find the equipment, raw materials and packaging to make the food in the required amounts. A labour force may need to be trained and finance obtained. Finding out about all these things is known as doing a **feasibility study**.

The general procedure for conducting feasibility studies is to:

1. gather information
2. analyse it
3. use the analysis to plan the business.

The feasibility study has three components (summarized opposite).

1.	Marketing aspects:	the demand for the food, how much to make each day, how to promote and sell the food
2.	Technical aspects:	including premises, equipment, raw materials, quality control and packaging needed to produce this amount of food
3.	Financial and legal aspects:	loans or finance needed to support this level of production, business registration, certificates, financial arrangements with suppliers and customers.

These aspects are elaborated in Chapter 5.

A feasibility study gives the entrepreneur the opportunity to think through what the business will involve in practice, to identify likely problems and to give him/her the confidence to go ahead.

When the results of the study are organized and written down, it is known as a **business plan**. If a loan is needed, the funding agency or bank will usually need a simple business plan to show that the entrepreneur is serious about the work and has thought about likely problems. This gives the funders more confidence that their loan will be repaid.

1st stage
Market feasibility
- Market research
- Selling strategy
- Expected market size/share
- Competitors

2nd stage
Technical feasibility
- Scale of production needed to meet market share
- Equipment, materials, services and labour needed for scale of production selected
- Quality control
- Distribution

3rd stage
Financial feasibility
- Startup costs
- Cashflow for one year (income and expenditure)
- Loan required
- Business development over three years
- Profitability/sustainability

- Decision

The three stages of a feasibility study

So the business plan (see page 42) has the following functions:

- It is the basis of a loan request
- It shows that the entrepreneur has thought seriously about the business and how it will develop
- It is a working tool to help the entrepreneur plan for the future.

THE DIFFERENT ASPECTS of a feasibility study are described in more detail below.

Sales and marketing

When the entrepreneur knows the scale of production needed to supply a known share of the market (Chapter 2), the next step is to decide in detail how the products will be sold (including the type of promotion, type of sales outlet, and so on) and how to deal with competitors. The maximum price that customers or retailers will pay for the product is also important for planning the finances of the business.

All products need promotion either to introduce them into the market or to increase the demand. This is achieved through the use of the media (radio, TV, newspapers, etc.) or by posters, personal and family contacts, discounts, new sales policies to retailers etc. – whichever is found to be most effective at the lowest cost.

The best form of promotion is a reputation for good products, friendly service, fair prices and good management. The cheapest form of promotion is recommendation by satisfied customers.

Advertise using newspapers

Posters may be less expensive than other ways of advertising

Samples for people to taste are a good way of promoting a product – especially a new product

Leaflets can be used to promote a product

Radio advertising can be very effective

Signboards given to retailers help to advertise your product

Good packaging and presentation of a range of products from the same manufacturer (Academy of Development Science, Kashele, India)

A note about packaging

Packaging not only protects the food against deterioration, but also has a very important marketing role. It encourages a customer to buy a product in preference to others alongside it on the retailer's shelf. It tells potential customers about the food before they have a chance to taste it and good packaging will imply a better quality product. It is therefore worthwhile making the effort to have the best packaging and labelling that can be afforded.

Technical requirements

A number of different technical inputs are needed to make a food in the required amount and to the expected quality and price. This needs careful thought and planning to ensure that all aspects of a process operate together without hold-ups, unnecessary expense or wastage.

The series of questions below is helpful in deciding the technical requirements of the business:

- Is a suitable building available? What modifications are needed?
- Are services (fuel, water, electricity, etc.) available and affordable?
- Are trained workers available and are their salaries affordable?
- Are enough raw materials available when needed and of the correct quality?
- Is the cost of the raw materials satisfactory for year-round production?
- Is the correct size and type of equipment available for the expected production level and at a reasonable cost? Can it be made by local workshops?
- Are maintenance and repair costs affordable?
- Is sufficient information and expertise available to ensure that the food is consistently made at the required quality?
- Are suitable packaging materials available and affordable?
- Are distribution procedures to retailers or other sellers established?

In all of these aspects it is best to seek the advice of a qualified food technologist to decide on the best method of production. For many foods this is essential to avoid the risk of food poisoning.

The building should be easy to keep clean and free from rodents

Are alternative fuels available?

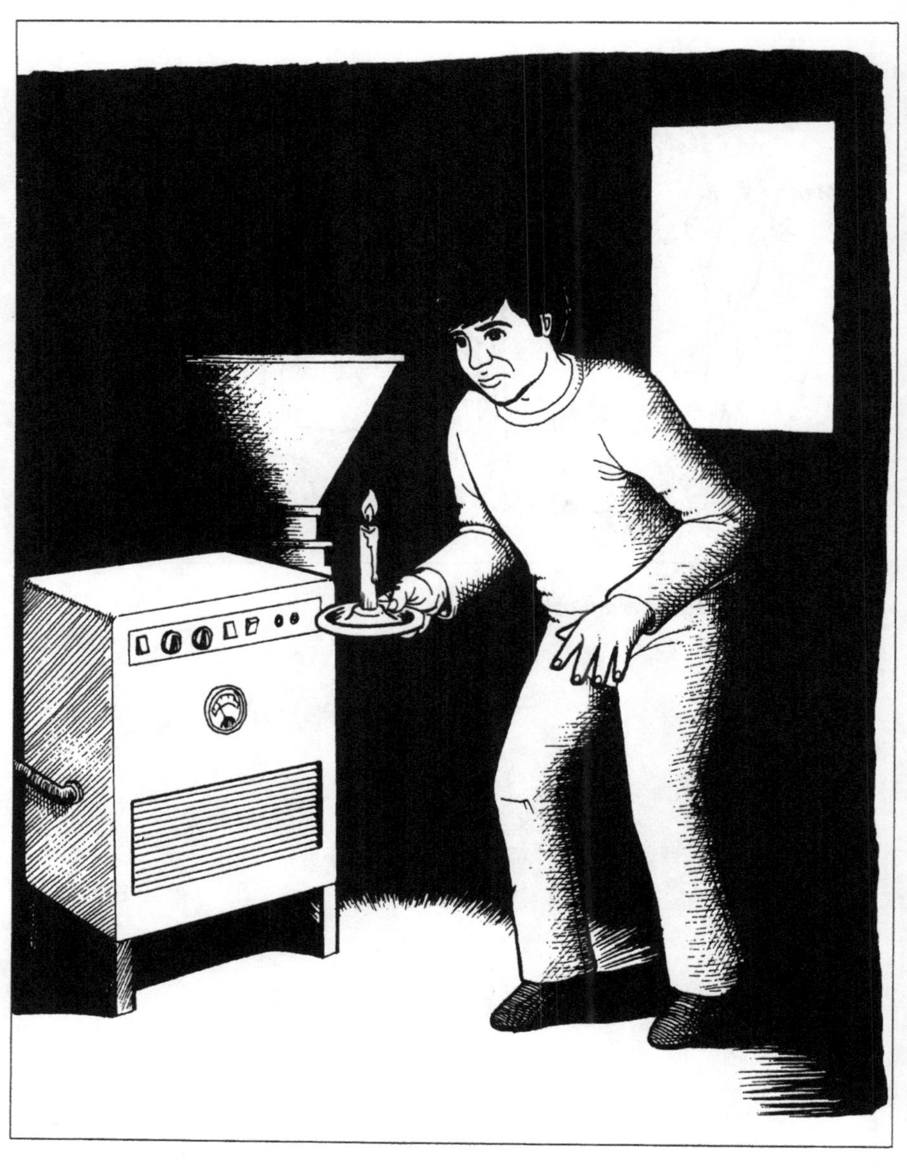

Is the electricity supply reliable?

Trained staff help to ensure good quality

Is the equipment too small or too large for the expected scale of production?

Will maintenance and repairs to equipment be too expensive?

Are distribution methods to customers or retailers already available?

Legal considerations

Registration of the business

The laws concerning the types of enterprise that can be set up vary in different countries. There are also advantages and disadvantages to different types of enterprise according to the legal and economic framework of the country concerned. The best structure for a business depends partly on the wishes of the owners and partly on the type of production to be undertaken, but in many countries it is easiest to start as an individual enterprise.

Professional advice on the best type of enterprise is needed from small enterprise advisers, accountants or solicitors before starting registration procedures.

Types of business

In many countries the law allows the following types of enterprises:

- business with limited liability (limited company with several directors)
- individual enterprise with limited liability (limited company with owner/director)
- personal enterprise (no limited liability)
- un-incorporated association (no limited liability)
- co-operative association
- not-for-profit organization
- registered charity (in some countries these are not allowed to trade).

Small businesses often begin with one person

If the demand for your products is large enough a bigger business can be started

Finance

Startup costs

'Startup capital' is the total cost involved in buying or converting a building, buying equipment, registering the business, training staff and buying packaging and initial raw materials. These should all be calculated to determine whether the owner's money (termed 'equity') will be sufficient to start the business. If not, a loan may be needed from a bank or other lender.

Do you have enough money for all the different expenses involved in starting a business?

Operating costs

The costs involved in producing a food can be divided into 'fixed costs' (those that have to be paid even if no food is produced) and 'variable costs' (those that change according to the amount of food that is produced). These costs should be calculated in advance assuming a scale of production from the share of market demand. If a loan is taken, the costs of repayment should be included in the fixed costs. A sample calculation is shown in Table 3.

Table 3: Daily production costs and income

For 200 pots of tomato sauce

	$	
Fixed costs		
Rent	2.0	
Labour (permanent + owner's salary)	15.0	
Insurance	0.3	
Professional fees	1.0	
Maintenance/repair (building and equipment)	2.6	
Loan repayment	1.3	
Interest charges	0.9	
Depreciation	0.4	
Business licences	0.2	
Sub-total		23.7
Variable costs		
Raw materials	18.6	
Other ingredients	6.6	
Packaging	22.0	
Transport	4.5	
Electricity	1.8	
Fuel	6.9	
Water	0.4	
Labour (temporary × 3)	18.0	
Advertising	7.6	
Sub-total		86.4
Total costs		110.1
Income (at $0.6 per pot)		120.0
Gross profit (income – costs)		9.9

When pricing a product, add up all the different costs, including distribution costs, and find out whether customers will pay this amount for your food

Price of the product

The correct price is important to be able to enter the market and to sell the product at a profit. To establish the right price it is necessary to consider the total costs involved in delivering the product to customers and the competitors' prices (note that this calculation should include the profit expected by the wholesaler or retailer).

The selling price should be continually reviewed as production costs may vary with, for example, availability of raw materials or labour, and variable customer demand. The price of the product should allow the entrepreneur and distributors to make an adequate profit and still have a good demand for the product.

The scale of production is initially estimated from information about the expected market share – sales. When the costs of production and expected income are calculated it is then necessary to see whether the 'breakeven point' has been achieved. This is the production level at which the total costs will equal the total income if everything produced is sold. Production levels should not fall below this point. It is possible to operate at this point for a short time without making a profit. Production levels must always be above this point for the business to be profitable and for workers and owners to receive an income.

To calculate the breakeven point of production, the costs need to be separated into *fixed* and *variable* costs (see page 35). The following equation can be used to determine the level of production required to break even.

$$\frac{\text{Fixed costs}}{\text{Revenue} - \text{Variable costs}} = \text{Production level at breakeven point}$$

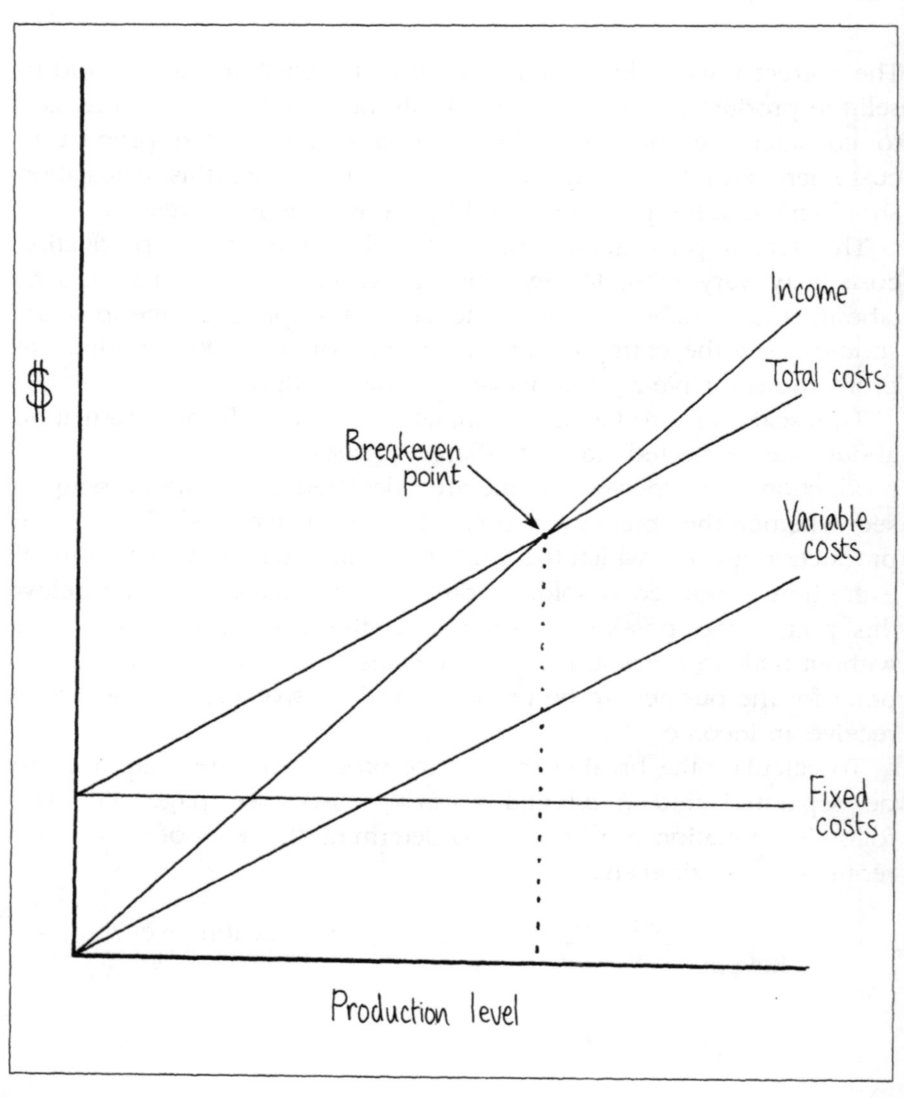

This is how income and costs change with different levels of production: the production level should always be more than the breakeven point

If the total cost of production is higher than the expected income, the business will make a loss. The entrepreneur may decide to forget the idea and start again with a different product, or see if production costs can be reduced by looking carefully at each of the main costs (such as raw materials or rent) to find areas where savings can be made.

When variable cost are low and the potential for selling more exists, it may be possible to become profitable by increasing production levels. Increasing the selling price is only viable if the product remains competitive.

If the break even production levels of the business look viable a 'profit and loss statement' and a 'cashflow forecast' are produced.

Profit and loss statement

A profit and loss statement shows income and expenditure over a fixed period. It shows money earned and used over that period. The following example gives figures for a five-year period.

Year	*1*	*2*	*3*	*4*	*5*
Gross income	21,300	22,330	26,550	33,400	34,000
Operating expenses	22,600	22,800	23,990	25,750	27,870
Gross profit/loss	(1,300)	(470)	2,560	7,650	6,130
Depreciation	800	800	800	800	800
Income before taxes	(2,100)	(1,270)	1,760	6,850	5,330
Corporate income tax @ 50 per cent	–	–	880	3,425	2,665

Figures in the second row are subtracted from those in the top row to give profit/loss each year. Interest payments (an expense) are included in the Operating expenses statement. Figures in brackets show a loss. Depreciation (wear and tear) is calculated by dividing the total equipment cost by the number of years it is expected to last.

CASHFLOW

MONTHS	JAN	FEB	MAR	APR	MAY	JUN	JUL	AUG	SEP	OCT	NOV	DEC	TOTAL
SALES INCOME	20		20		50	20		20	50	20		20	220
EXPENSES	10	10	10	10	10	10	10	10	10	10	10	10	120
ACCUMULATED PROFIT OR LOSS	10	0	10	0	40	50	40	50	90	100	90	100	100

A cashflow forecast shows the amount of money the business will actually have and it can be used to predict when a second instalment of a loan may be needed (for example, in February or April)

Cashflow forecast

A cashflow forecast is essential when starting a new business or expanding an existing one, to enable the entrepreneur to plan ahead. A table can be compiled, showing sales incomes and expenses on a monthly basis. The monthly profit or loss is calculated by subtracting the expenses from the income.

From this information the entrepreneur will see when there are profitable months or when a loss is expected. For example, sales may fall at certain times of the year and the price of raw materials may increase, especially just before the harvest season.

Once the forecast is made for the first year, a similar one is made for the next two years to reflect the expected development of the business. This should take into account increases in price, changes in sales and the action of competitors.

YEAR II

MONTH	1ST	2ND	3RD	4TH	5TH	6TH	7TH	8TH	9TH	10TH	11TH	12TH	TOTALS
SALES INCOME	2666	2666	2666	-0-	2666	2666	-0-	2666	2666	2666	-0-	-0-	21,328
EXPENSES	2599	2599	2449	160	2449	2449	160	2449	2449	2449	160	160	20,532
PROFIT OR LOSS EACH MONTH	67	67	217	(160)	217	217	(160)	217	217	217	(160)	(160)	796
ACCUMULATED PROFIT OR LOSS	4543	4610	4827	4667	4884	5101	4941	5158	5375	5592	5432	5272	5,272

YEAR III

MONTH	1ST	2ND	3RD	4TH	5TH	6TH	7TH	8TH	9TH	10TH	11TH	12TH	TOTALS
SALES INCOME	2666	2666	2666	-0-	2666	2666	-0-	2666	2666	2666	-0-	-0-	21,328
EXPENSES	2449	2449	2449	160	2449	2449	160	2449	2449	2449	160	160	20,232
PROFIT OR LOSS EACH MONTH	217	217	217	(160)	217	217	(160)	217	217	217	(160)	(160)	1096
ACCUMULATED PROFIT OR LOSS	5489	5706	5923	5763	5980	6197	6037	6254	6471	6688	6528	6368	6,368

Cashflow forecasts rarely involve simple arithmetic like that shown opposite and are more likely to resemble those shown here

Preparing a business plan

Preparing a business plan helps the entrepreneur to clarify and focus his/her ideas and to make the mistakes on paper rather than in the operation of the business. When the plan is complete and shows that a successful business is possible, it makes the entrepreneur feel more confident about success. It also helps to make it clear how much money is needed and, if properly prepared, it will give the banker or other loan agency confidence that their money will be repaid.

The main considerations when preparing a business plan are:

- make the plan as easy to understand as possible by using simple language (lenders rarely understand food processing)
- include as much detail as possible and if necessary do thorough research first
- look outwards from the business to judge what competitors will do and how the business will develop to become sustainable.

The information in the business plan will help the entrepreneur to make decisions on:

- whether the business is a good idea: will it work?
- the demand for the product
- the resources available to produce the product at the right quality and for the right price
- whether the business will be profitable
- whether a loan is needed, and if so, how much and when.

How to write a business plan is described in Appendix 2 and examples are given in Appendix 3.

A well-prepared business plan will help to get a loan

After conducting a feasibility study and preparing a business plan, the entrepreneur should arrange for adequate finance to be made available when it will be needed. He/she can then begin the detailed process of actually getting production started.

Producing foods for sale involves a different set of problems to producing them in the home, the most important being to ensure a consistent quality of product made at the lowest achievable cost.

Registration and certificates

In most countries a new business must be registered with the local government authority, especially in urban areas, and the taxation authorities. Most countries also require a food business to be inspected and authorized, often needing a certificate of hygiene or similar documentation before production can begin. Sometimes a certificate is needed from the Bureau of Standards, Ministry of Health or similar authority to show that the food products have been analysed and found to be safe. The authorities may also require the foods to conform to legal standards regarding composition. In all cases professional advice from a food technologist is needed before starting production.

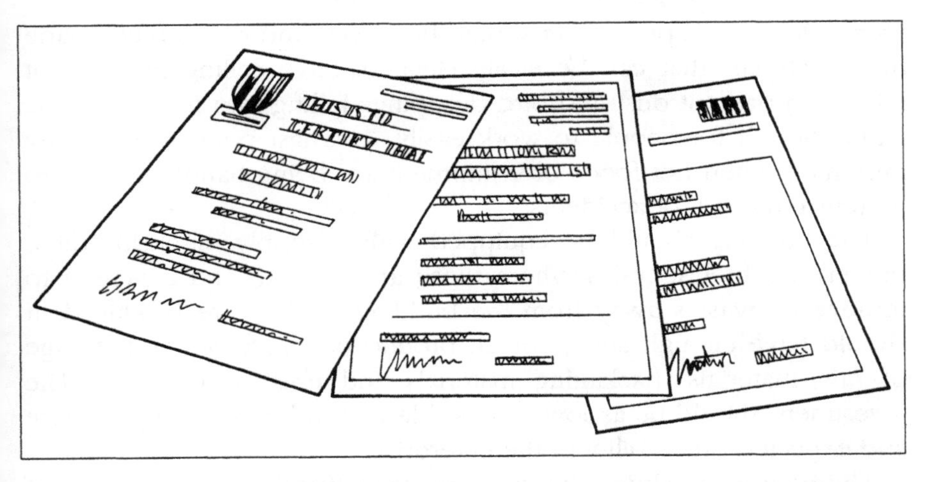

Hygiene certificates and registration documents are usually needed before a food business can begin production

Premises and production facilities

Premises for food processing should be insect- and rat-proofed, made from materials that can be easily cleaned, have ceiling panels, not rafters, to prevent dust and bird droppings falling into foods, and be large enough for people to work easily. This last point is especially important when hot foods or equipment are being handled, so as to minimize the risk of accidents.

The building should be equipped with a supply of clean water, power and fuel supplies where these are needed, and a system to remove all wastes away from the building as they are produced. It should have enough space for all production processes and storage of raw materials, packaging materials and finished products. The investment should be as low as possible and in keeping with the size and expected profitability of the enterprise.

The equipment should be the correct size for the intended scale of production and should be laid out in the room to minimize the risk of cross-contamination. Regular maintenance and cleaning schedules should be designed and managers should ensure that they are followed.

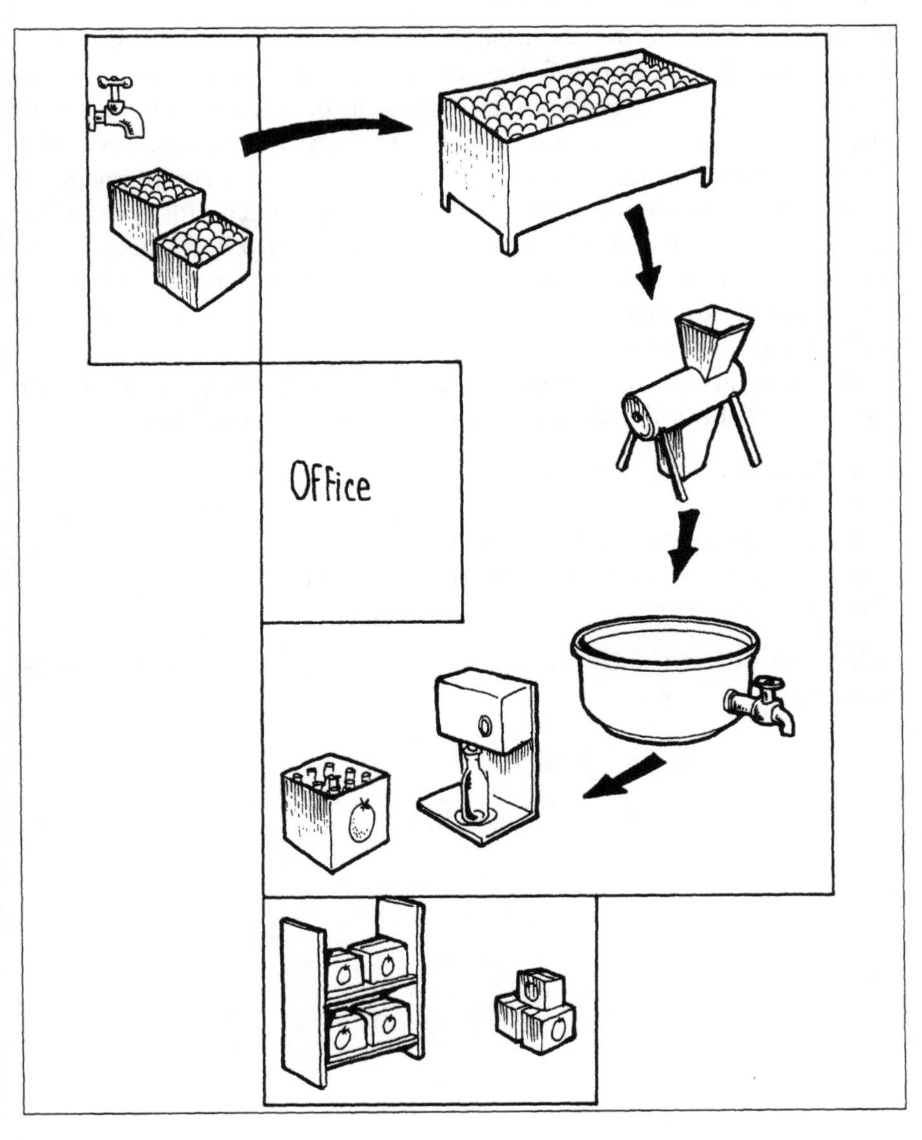

Arrange the equipment to allow easy working and avoid cross-contamination between raw materials and finished products

Packaging and promotion

At an early stage in the development of the business it is necessary to decide on the design of the packaging and promotion of the products. A first step is to design a symbol that clearly identifies your products and establishes the differences from the competitors. This 'logo' can then be used on all products to develop a 'brand image'. It is strongly recommended that a specialist commercial artist or graphic designer is employed to prepare the final artwork for labels and promotional materials. Negotiate with printers the cost of supplying the required amount and see examples of their previous work to judge the quality of printing.

The law in most countries states that labels must at least give information about the company and the product, as follows:

- the name of the product
- the net weight
- the ingredients in order of amount
- the name and address of the manufacturer
- the brand name.

Other information such as 'use by' or 'sell by' dates may also be required.

A 'generic' label shows details of the company; the names of different products in a range can be stamped on to the label as they are produced

Supplies

Make arrangements with suppliers of raw materials, packaging materials and services to supply the correct amounts at the right time. Wherever possible, arrange credit facilities with the suppliers. Costs can be reduced by contracting farmers for raw materials rather than buying them from local markets.

Once the production cost structure is known, it is possible to decrease the amount or cost of some ingredients or raw materials. This is done by identifying the principal raw materials that have the largest effect on costs, then trying to reduce the amount or get them more cheaply.

To increase profits, there are two alternatives:

- increasing income through expanding sales or increasing the price of the product

or

- reducing production and distribution costs.

For both alternatives, it is necessary to study what is possible and take careful decisions on changes to production, administration and marketing.

Too many labels and not enough bottles is an example of poor production planning

Internal organization and staff training

The entrepreneur cannot do all the jobs within the business (although many will try to do this when they start), and some form of sharing the responsibilities is needed. The entrepreneur should employ people who have the skills that are needed. For example, a trusted person who is able to keep records and look after financial aspects; a person who is careful and meticulous to check that quality assurance procedures are being followed and that product quality is uniform; thoughtful and innovative production workers who will think about improvements to their work and follow production schedules accurately.

A system of checks is usually needed to ensure that a single employee does not have complete responsibility for a whole area of the business (for example, records of buying and using materials should be kept by the person responsible for the accounts and not by the storeroom worker).

The entrepreneur needs to decide the internal structure, the levels of authority in the enterprise (who decides what, how decisions are made), the different responsibilities for production, marketing and administration, and the task each person must perform. The entrepreneur is also responsible for ensuring that each person is properly trained to do their job efficiently and without endangering their own health and safety or that of other people working with them.

A well organized business is more profitable

Quality assurance

The quality of food products is the most important characteristic after price. If quality varies, customers will not know what to expect and as a result they will stop buying it. Quality assurance is a management system that is used to ensure a *consistent* product every time.

(*Note:* the actual quality of a product is decided by the manufacturer to meet the needs of the customers at a price that they can afford.)

Successful quality assurance depends first on establishing detailed standards for hygiene and processing conditions and then ensuring that *everyone* involved in the business is trained to meet those standards every day. It is likely that entrepreneurs will need the advice of a food technologist to establish the detailed standards.

Distribution and selling

Transport, distribution and marketing costs are often a major expense to small businesses. It is also true that if products are not available when and where customers wish to buy them, then sales will not take place. The entrepreneur has to ensure a distribution system that will enable products to reach customers on time and in the correct condition as economically as possible.

The advantages and problems of dealing directly with the final consumers should be compared to working with other enterprises, traders, retailers or intermediaries to get products to consumers. A payment system that suits the producer, middlemen and customers should be established (for example cash in advance, by consignment or credit).

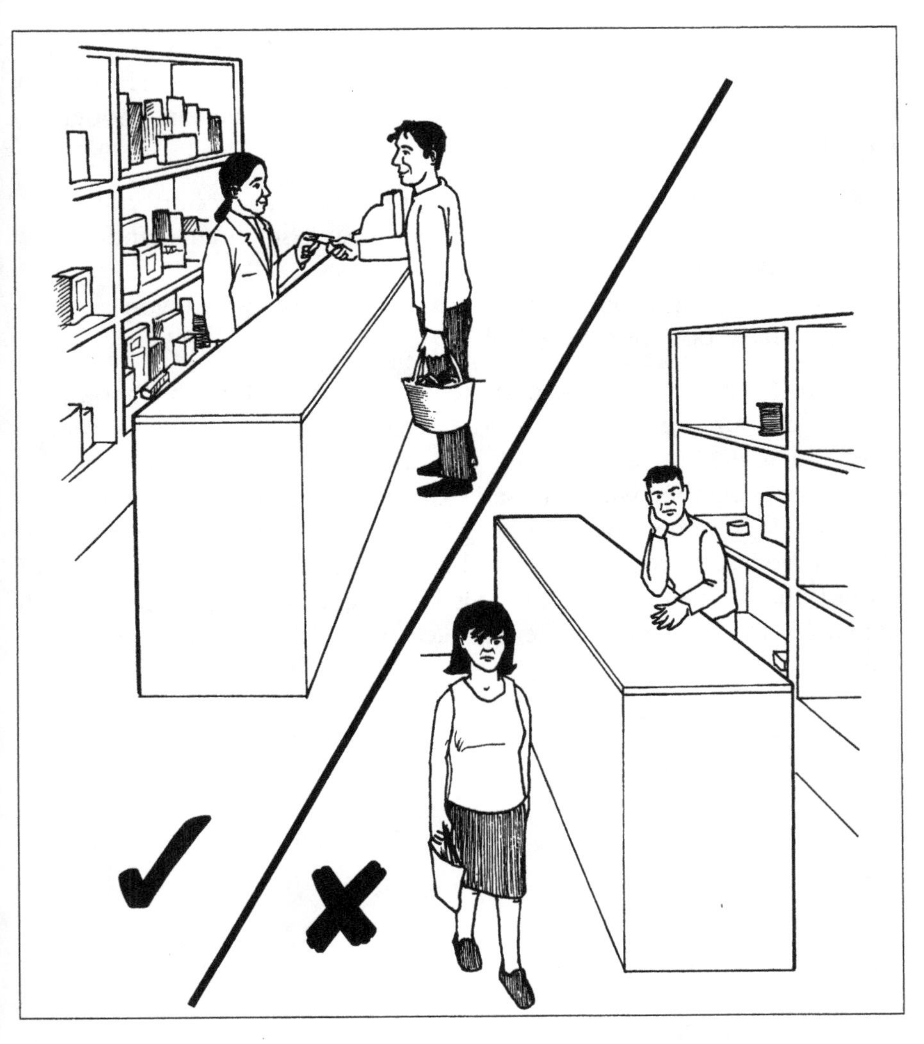

If you do not make sure that your products are always available in shops there will be nothing for customers to buy

Record keeping

An accountant may be needed for larger businesses to reduce the work involved in preparing accounts and to reduce the amount of taxes paid. However, for smaller businesses it is usually enough to have the following record books:

- For recording money: cash book, sales book, account book (or ledgers for transactions and salaries), bank statements and cheque book.
- For recording production details: stock book, production book, purchase book, delivery book.

A small enterprise can use an account book and a cash book as a simple accounting system. In the account book the entrepreneur keeps a daily register of all the income and expenses of the enterprise, together with any sales or purchases using credit. It does not matter whether the money goes in or out but the transaction should be registered. The cash book registers just the amount of money that comes in and out. This book records the amount of cash available to buy raw materials, to pay bills or other business expenses.

Keep your records and accounts up to date to avoid confusion and loss of money

Taxation

Each country has its own tax system. In many countries there are the following taxes:

- Company's Income tax
- Business tax
- Sales tax
- Excise tax
- Social Security contribution
- other contributions to pensions/employee benefits

It is necessary to register with the taxation authorities in order to avoid prosecution later on and possible closure of the business.

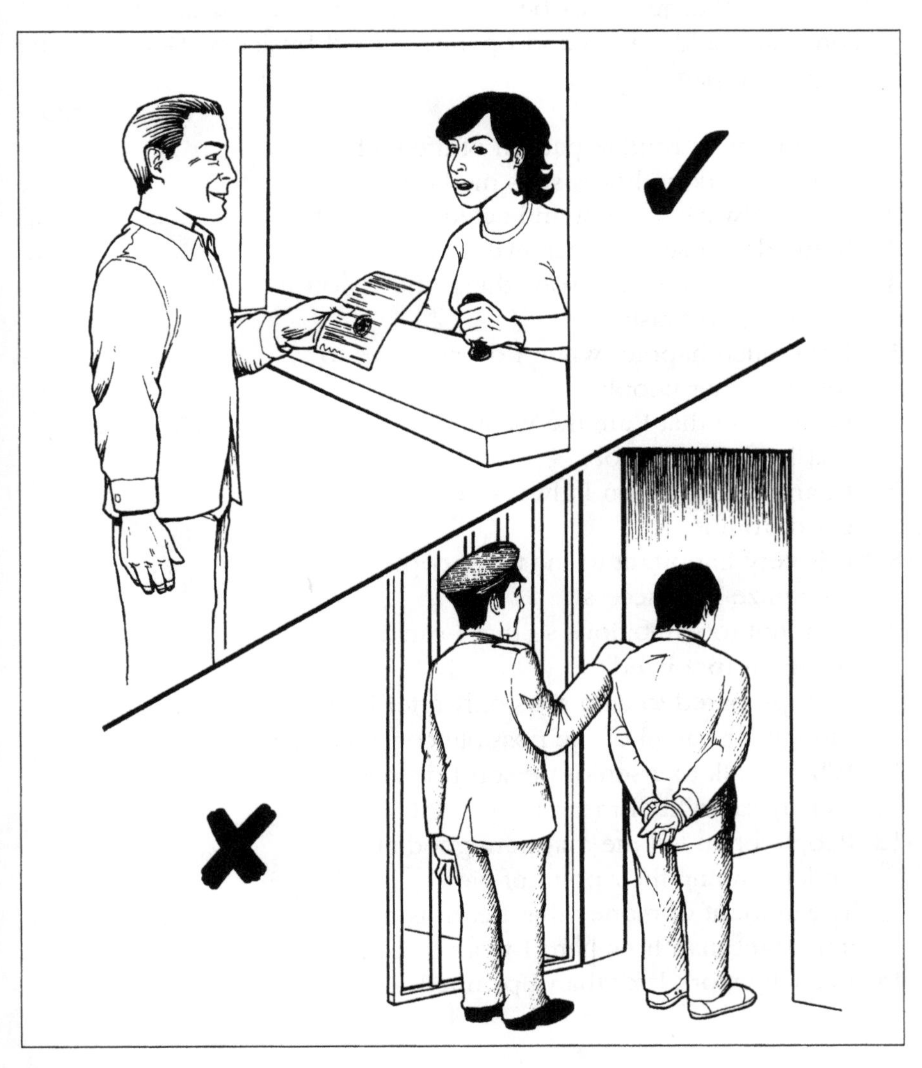

A receipt for payment of taxes helps the business to develop without problems from the authorities; failure to pay business taxes can result in closure of your business and fines or imprisonment

Ask yourself the questions below to see whether you are the sort of person that could set up and operate a food business. Details of the scores are on page 123.

		YES	NO
1	I think that a routine pattern of life with regular working hours suits me best	....	....
2	I have always thought and acted by myself	....	....
3	Some days I seem to achieve nothing	....	....
4	It is not good to start something unless you are going to finish it	....	....
5	I am much happier when I do not have to rely on other people	....	....
6	I often feel that I am the victim of events that I cannot control	....	....
7	In any bad situation I always get something good from it	....	....
8	It is very important to me that people recognize my success	....	....
9	I am not too ambitious so that I can avoid being disappointed	....	....
10	I am prepared to take risks only after I have thought about all of the possible consequences	....	....
11	When I talk to a senior person I do not usually say what I mean	....	....
12	People often tell me that I am good at understanding their point of view	....	....
13	The amount of money that I earn is more important than how hard I work to earn it	....	....
14	I usually work later than I planned	....	....

The information collected by the entrepreneur needs to be written down in a simple, concise way to show bankers or other lending agencies that the business is carefully planned, in order to get a loan. There is no fixed way of doing this, but the example below has been used successfully.

1. *Start with an introduction*
This summarizes what the product is, why the business is a good idea and who the customers are.

2. *Give some basic information*
For example the name and address of the business, the owners and workers, their qualifications and experience.

3. *Describe the product*
Give details of the raw materials, the production processes, quality control checks, packaging etc. (If you can take samples to the lender this will help.) Say what is special about your product.

4. *Describe the market*
Who are the customers? Where are they? How big is the market (both size and value)? Is demand growing or falling? Who are the competitors and what will they do if you start production? What are their strengths and weaknesses? What is the value of the market over a year? What is your market share?

5. *Describe your selling plan*
How you will distribute and sell your product, what promotion you will use, what the competitors do, what the product cost will be, why your methods will be successful.

6. *Describe the premises/equipment needed*
Where the business will be located and why, what sort of building you will use, whether it meets health and hygiene laws, what services are needed, what equipment you must buy/make and what are the

costs. Do not forget storage and distribution.

7. *Describe the finance needed*
What finance will you need to start the business and operate it for one year (include profit and loss and cash flow)? How much of your own resources will be put into the business? What size of loan is needed and what it is needed for? What security will you be able to offer on the loan?

8. *What are your plans for the future?*
What are your objectives in running the business? How will you achieve them? What do you expect to happen over the next three to five years? (Include cashflow forecast.)

Business plan 1: Coconut ice sweets

Overall assessment

The business plan is well-written, concise and clearly thought through, especially the cashflow. More detail is needed in some sections to give a better impression of what is planned. The financers feel that a one-year loan is too short because the business is planned to become profitable in the second year. The plan is essentially a static view of the business and contains little on future development and expansion which would help to show that it is a sustainable venture.

Coconut ice processing as small-scale on the coast of Perú

Diana Colquichagua R.

Summary

This project is a small business proposal for the making of coconut ice with a production of 5580 kg per year, around the city of Lima, the capital of Perú.

In the first year of the business, it will concentrate on the establishment of the market and the trade mark.
In the second year, it will develop other confectionary products using the same equipment and will increase its production.

The principal target customers of the enterprise will be children.

Desiccated coconut and sugar are the principal ingredients, which are available in Lima.

The "coconut ice" is a kind of confectionary product where the principal ingredients ie the sugar and glucose syrup give the special textural characteristics. The flavour depends on the coconut.

A clear, concisely written introduction which sets out the main points of the proposed business.

Coconut ice can be of various flavours and colours. Its shape can be cubes of pink and white, yellow or orange and white, or in some cases, a combination of three colours.

Three cubes will be wrapped together with cellophane paper and then tied on both ends with coloured string. Each unit will weigh 75g.

The product will be distributed to bakeries and small shops in boxes of ten units and the rest of the product will be sold directly through sellers.

This small enterprise will produce 62,400 units of coconut ice per year and it will give employment to two permanent workers, one supervisor and three direct sales persons.

The capital required to start the enterprise is US$ 2059·40 of which the owner will put in 22% of his own money ($ 459·40). The rest will be financed by Commercial Bank ($ 1600).

The Production Process

Initially, the enterprise will employ one supervisor and two workers to set up the factories.

The manufacturing process of coconut ice only requires simple equipment. The list of equipment is shown in Annex I. The raw materials required are also listed in this Annex.

Coconut ice is made by the following process:

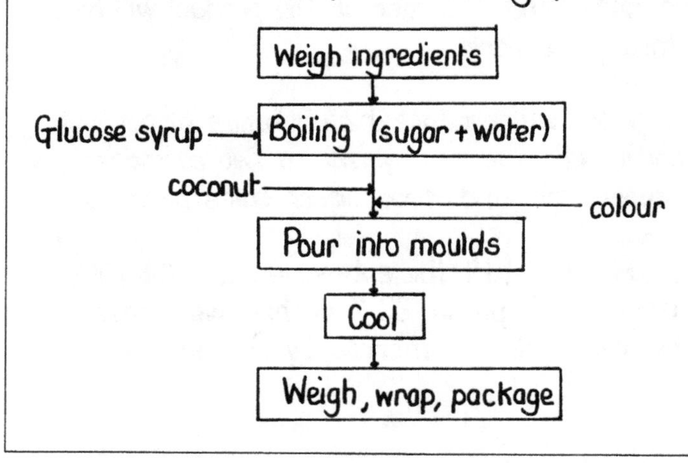

It is useful to have details of the production process as these are often missing from business plans. (Note: Annexes are not reproduced in full.)

Market Share

It is estimated that there are 100 schools near the factory and there are about 200 students in each school. It is estimated that 0·8→1 % of students will buy the product every day.

total sales per day : 100 x 200 x 1% = 200 units
total sales per day : 100 x 200 x 0·8% = 160 units

There are 36 bakeries, small shops and tea rooms located near the factories. 26 of these outlets are expected to sell between 1·5 to 2·5 units per day.

total sales per day : 26 x 2·5 = 65 units
total sales per day : 26 x 1·5 = 39 units

The total production per year is 62,400 units
The production plan is as follows:

(START PRODUCTION)	March → May	15kg/day	= 4,800 unit/month
	June → October	20kg/day	= 6,400 unit/month
	November → December	15kg/day	= 4,800 unit/month
	January → February	10kg/day	= 3,200 unit/month

Well-thought-out market research of defined customers, but more details are needed of the selling strategy.

The production plan above is based on the consumption pattern of the customers in Perú. The peak production from June to October coincides with the school term and the winter season when consumption of coconut ice is likely to be highest. There is a low demand for this product during the summer months of January and February.
Production will increase in the second year to include making other confectionary products.

Capital Required

The enterprise requires a capital outlay of US $ 2059. The owner is expected to put in US $459·00 of his own money. This represents 22% of the total capital. The rest will be financed by Commercial Bank.

The cash flow of the enterprise for the first year is shown in Annex 2. The profit budget is shown in Annex 3.

Financial Costs

Capital required to start the enterprise.

Fixed Costs

				$	
1.	Building				260-00
2.	Equipment -	weighing scales	(1unit)	173-00	
		cooking rings	(1 ")	185-00	
		saucepan (50lbs)	(2 ")	45-00	
		wooden spoon	(2 ")	1-00	
		thermometers	(2 ")	20-00	
		moulding tins	(4 ")	31-00	
		Knives	(3 ")	6-00	
		rulers	(1 ")	1-00	
		calculator	(1 ")	7-50	
		table & others	(1 ")	60-00	
		teaspoon	(2 ")	0-60	
					530-10
3.	Insurance		(3)	6-60	19-80
4.	Depreciation				2-10
5.	Rent		(2month)	45-00	90-00
6.	Salary		(1month)	100-00	100-00
7.	Administration		(1 month)	30-00	30-00
					$1032-00

A comprehensive list of equipment and materials. Hand-wrapping of the sweets may take longer or need more people than planned. Careful attention will be needed to schedule the work of the production staff properly.

Variable Costs.

1. Raw Material (for 1 month)	Kg	$/kg	$
- desiccated coconut	72	2·00	144·00
- sugar	288	0·80	230·40
- liquid glucose	2	3·20	6·40
- food colouring	250ml	30·00	7·50
2. Packaging Material	pieces	$/1000	$
- impression	5000	3·00	15·00
- boxes	1000	4·00	4·00
- cellophane	1 roll		22·00
- coloured string	6 metres		3·00
3. Promotion	$ 2·5 (3) (24)		180·00
4. Wages (workers)	$ 2·5 (24) (2)		120·00
5. Fuel (Kerosene)	$ 2·0 (24)		48·00
6. Transport	$ 4·0 (24)		96·00
7. Maintenance/repairs	$ 2·08		2·08
8. Power (electricity e water)	$ 9·00 (month)		9·00
			$ 887·00

Contingencies	5% fixed costs	51·6
	10% variable costs	88·7
	total	$ 2059·4
	owner →	459·4
	capital required →	$ 1600·00

Annex IA

Calculation

- To obtain the price of the product
- Consider 24 days of work per month and 288 per year

Fixed Costs		$
1. Building		260·00
2. Insurance	$ 20/month	240·00
3. Depreciation		20·00
4. Rent	$ 45/month	540·00
5. Salary	$ 100/month	1200·00
6. Interest		210·00
7. Administration	$ 30/month	360·00
		2830·00

The costings and cashflows on the following pages are well prepared. In particular the variations in seasonal demand are well reflected in the cashflow (although not in transport and administration which may also vary) and there is a three-month hold on repayments.

Variable Costs

			$
8.	Raw material - packaging, Fuel		6,243·90
9.	Promotion	$180/month	2,160·00
10.	Wages	$ 120/month	1,440·00
11.	Transport	$ 72/month	864·00
12.	Maintenance/repairs	$ 2·08/month	24·00
13.	Power	$ 9/month	108·00
			10,839·90

Total cost of production	13,669·90
Total number of units	62,400
Cost per unit	$0·219

∴ Selling price per unit is $0·30

Cashflow forecast

	MONTH March		MONTH April		MONTH May		MONTH June		MONTH July		MONTH Aug		TOTALS	
RECEIPTS	BUDGET	ACTUAL	BUDGET	ACTUAL	BUDGET	ACTUAL	BUDGET	ACTUAL	BUDGET	ACTUAL	BUDGET	ACTUAL	BUDGET	ACTUAL
Cash Sales	360.00		360.00		360.00		1,600.00		1,600.00		1,600.00		5,880.00	
Cash from Debtors	—		1,080.00		1,080.00		1,080.00		4,800.00		4,800.00		12,840.00	
Capital Introduced	2,059.4		—		—		—		—		—		2,059.4	
TOTAL RECEIPTS (a)	2,419.4		1440.00		1,440.00		2,680.00		6,400.00		6,400.00		20,719.4	
PAYMENTS														
~~Payments to Creditors~~ Raw Material	388.3		388.3		388.3		517.73		517.73		517.73		2,718.09	
Salaries/Wages	220.		220.		220.		220.		220.		220		1320.	
Rent/Rates/~~Water~~	45.00		45.00		45.00		45.00		45.00		45.00		270.	
Insurance + Packaging	64.00		64.00		64.00		78.6		78.6		78.6		427.8	
Repairs/Renewals \ Maintenance	24.00		24.00		24.00		24.00		24.00		24.00		144.	
Heat/Light/Power + water	9.00		9.00		9.00		9.00		9.00		9.00		54.	
~~Postages~~														
Printing/Stationery	180.		180.00		180.00		180.		180.00		180.00		1080.0	
Transport	96.00		96.00		96.00		96.00		96.00		96.00		576.	
Telephone, Admin	30.00		30.00		30.00		30.00		30.00		30.00		180.	
~~Professional Fees~~														
Capital Payments	772.10		—		—		—		—		—		772.10	
Interest Charges	—		—		—		25		25		25		75.00	
Other Fuel (Kerosene)	48.		48.		48.0		48		48		48		288.00	
~~V.A.T. payable (refund)~~ Loan Repayment	—		—		—		133.33		133.33		133.33		400.00	
TOTAL PAYMENTS (b)	1876.4		1100.3		1100.3		1406.66		1406.66		1406.66		8301.99	
NET CASHFLOW (a-b)	543.00		339.7		339.7		1273.34		4993.34		4993.34		12,482.42	
OPENING BANK BALANCE	0		543.0		882.7		1222.4		2495.74		7489.08		12,632.92	
CLOSING BANK BALANCE	543.10		882.7		1222.4		2495.74		7489.08		12,482.42		25,115.34	

Cashflow forecast

RECEIPTS	MONTH Set		MONTH Oct		MONTH Nov		MONTH Dec.		MONTH Jan		MONTH Feb.		TOTALS	
	BUDGET	ACTUAL	BUDGET	ACTUAL	BUDGET	ACTUAL	BUDGET	ACTUAL	BUDGET	ACTUAL	BUDGET	ACTUAL	BUDGET	ACTUAL
Cash Sales	1,600.		1,600.		360		360		800		800		5,520	
Cash from Debtors	4,800.00		4,800.		4,800.		1,080		1,080		2 400		18,960	
Capital Introduced	—													
TOTAL RECEIPTS (a)	6,400.		6,400		5,160		1,440		1,880		3,200		24,480	
PAYMENTS														
~~Payments to Creditors~~ Raw Mat	517.73		517.73		388.3		388.3		258.86		258.86		2329.79	
Salaries/Wages	220?		220.		220.0		220.		220		220		1,320	
Rent/Rates/~~Water~~	45.0		45.0		45.0		45.0		45.0		45.0		270.	
Insurance + Packaging	78.6		78.6		64.0		64.0		39.3		39.3		363.8	
Repairs/Renewals) Maintenance	24.0		24.0		24.		24.		24.		24.		144.	
Heat/Light/Power + Water	9.00		9.00		9.0		9.0		9.0		9.0		54	
Postages														
Printing/Stationery	180		110.0		180.0		180.		180.		180		1080.	
Transport	96		96		96		96		96		96		576	
Telephone , Adm.	30.		30.		30.		30		30		30		180	
Professional Fees														
Capital Payments														
Interest Charges	25		25		25		25		25		25		150	
Other , Fuel (kerosene)	48		48		48		48		48		48		288	
~~V.A.T. payable (refund)~~ Loan Repayment	133.33		133.33		133.33		133.33		133.33		133.33		800	
TOTAL PAYMENTS (b)	1,406.66		1406.66		1262.63		1,262.63		1,108.46		1,108.46		7555.5	
NET CASHFLOW (a-b)	4,993.34		4,993.34		3,897.37		177.37		771.54		2,091.54		16,923.68	
OPENING BANK BALANCE	12,482.42		17,475.76		22,469.1		26,366.47		26,543.84		27,315.38		12,482.42	
CLOSING BANK BALANCE	17,475.76		22,469.1		26,366.47		26,543.84		27,315.38		29,406.92		29,406.92	

Profit budget

	MONTH March		MONTH Ap		MONTH May		MONTH Jun		MONTH Jul		MONTH Ag		TOTALS	
	BUDGET	ACTUAL	BUDGET	ACTUAL	BUDGET	ACTUAL	BUDGET	ACTUAL	BUDGET	ACTUAL	BUDGET	ACTUAL	BUDGET	ACTUAL
SALES (a)	1440.0		1440.0		1440.00		2,880-		6,400.00		6,400.00			
Less: Direct Costs														
Cost of Materials	480.3		480.3		480.3		624.33		624.33		624.33			
Wages	120.00		120.00		120.00		120.00		120.00		120.00			
GROSS PROFIT (b)	839.7		839.7		839.7		1935.67		5,655.67		5,655.67			
Gross Profit Margin (%, x 100%)	58..		58.31		58.31		72.2%		88.6		88.6			
Overheads														
Salaries	100.00		100.00		100.00		100.00		100.00		100.00			
Rent/Rates/Water	45.00		45.00		45.00		45.00		45.00		45.00			
Insurance	20.00		20.00		20.00		20.00		20.00		20.00			
Repairs/Renewals / Maintenance	24.00		24.00		24.00		24.00		24.00		24.00			
Heat/Light/Power + Water	9.00		9.00		9.00		9.00		9.00		9.00			
Postages														
Printing/Stationery	180.00		180.00		180.00		180.00		180.00		180.00			
Transport	96.00		96.00		96.00		96.00		96.00		96.00			
Telephone administration	30.00		30.00		30.00		30.00		30.00		30.00			
~~Professional Fees~~														
Interest Charges							25.00		25.00		25.00			
Other														
TOTAL OVERHEADS (c)	504		504		504		529.		529		529			
TRADING PROFIT (b)-(c)	335.7		335.7		335.7		1406.67		5,126.67		5,126.67			
Less: Depreciation	1.6		1.6		1.6		1.6		1.6		1.6			
NET PROFIT BEFORE TAX	334.1		334.1		334.1		1405.07		5,125.0		5,125.0			

Profit budget

	MONTH Sept		MONTH Oct		MONTH Nov		MONTH Dec		MONTH Jan.		MONTH Feb		TOTALS	
	BUDGET	ACTUAL	BUDGET	ACTUAL	BUDGET	ACTUAL	BUDGET	ACTUAL	BUDGET	ACTUAL	BUDGET	ACTUAL	BUDGET	ACTUAL
SALES (a)	6400.00		6,400.		5,760.00		1440		1880		3200			
Less: Direct Costs														
Cost of Materials	624.33		624.3		480.3		480.3		326.16		326.16			
Wages	120.00		120.		120.00		120.00		120.00		120.00			
GROSS PROFIT (b)	5,655.		5,655		4,559.7		839.7		1433.84		2,753.84			
Gross Profit Margin (b/a x 100%)	88.6		88.6		88.3		58.3		76%		86%			
Overheads														
Salaries	100		100		100		100		100		100			
Rent/Rates/Water	45.00		45.00		45.00		45.00		45.00		45.00			
Insurance	20.00		20.00		20.00		20.00		20.00		20.00			
Repairs/Renewals) Maintenance	24.00		24.00		24.00		24.00		24.00		24.00			
Heat/Light/Power + Water	9.00		9.00		9.00		9.00		9.00		9.00			
Postages														
Printing/Stationery	180.00		180.00		180.00		180.00		180.00		180.00			
Transport	96.00		96.00		96.00		96.00		96.00		96.00			
Telephone , adm.	30.00		30.00		30.00		30.00		30.00		30.00			
Professional Fees														
Interest Charges	25.00		25.00		25.00		25.00		25.00		25.00			
Other														
TOTAL OVERHEADS (c)	529		529		529.		529.00		529.00		529.00			
TRADING PROFIT (b) - (c)	5,126.67		5,126.67		4030.7		310.7		904.84		2224.84			
Less: Depreciation	1.6		1.6		1.6		1.6		1.6		1.6			
NET PROFIT BEFORE TAX	5,125.0		5,125.0		4029.1		309.1		903.24		2223.24			

Business plan 2: Peanut–popcorn snack food

Overall assessment
This is a very well-prepared plan with much attention to detail and careful thought about business planning. The phases of the business plan, the realistic sales projections and gradual expansion all give us confidence in its success and hence a willingness to support the venture. We would also like to see more details about the possible threat from competitors in the second year and more details about the process and quality control to be used. Despite this, the plan would be favourably considered for funding.

BUSINESS PLAN

M'S FOOD PROCESSORS
P.O. BOX MR 301
MARLBOROUGH
HARARE, ZIMBABWE

3 SULGRAVE ROAD
MARLBOROUGH
HARARE, ZIMBABWE

TEL: 39899

1.	INTRODUCTION
	An emergent business enterprise is introducing a new product: POPCORN WITH CRUNCHY PEANUTS.
2.	DETAILS OF THE BUSINESS
2·1	NAME: M's FOOD PROCESSORS
2·2	TYPE: Partnership (Family members)
2·3	ADDRESS: Postal: P.O. Box MR 301, Marlborough, Harare. Physical: 3 Sulgrave Road, Marlborough.
2·4	TELEPHONE: (Current) 39899
2·5	PERSONAL DETAILS
2·5·1	Bertha MSORA (Manager) - Adult Educator - Sociologist. Specialized in personnel administration. Also studied business administration/management. - Promoter of entrepreneurial programmes (especially for women) - Attended 3-month course on food processing: Silsoe College, U.K.

This introduction should make it clear that the popcorn is ***coated*** *in small peanut pieces. Full details of business name, address and personnel are well presented. It is important for potential funders to have full information and this outline is good.*

2.5.2	Addison MSORA (Salesman) – 5 'O' levels. 1 'A' level – Experienced in salesmanship – Worked at supervisory level in a hotel/restaurant. – Dormant partner in a catering enterprise
2.5.3	Joseph MSORA (To be trained as book-keeper) – 5 'O' levels No relevant experience.
2.5.4	Labourers : To hire three labourers.
3	THE PRODUCT POPCORN : with crunchy peanuts. This is popped corn coated with sugar syrup and covered with roasted peanut pieces.

The details of staff experience enable funders to assess their abilities and also to feel more involved in the business.

	<u>Major Influencing Factors for Choice of Product</u> · Easy availability of raw material (popcorn, sugar, oil, peanuts) · Simple technology · Easy to process · Nutritional New product – not offered by competition. · Target consumer – children · Affordable · Profitable
4	THE MARKET
4·1	<u>Target market</u>: 100 primary schools in Harare within a radius of 40 kilometres from location of enterprise.
4·2	6% of this market covered in the pre-establishment / feasibility study phase (see section 7)
4·3	15% to be covered in PHASE II of the business plan over the next year.

This section on influencing factors shows that the applicant has thought about the business in some depth. This gives confidence to potential funders. The section on 'market' is very important. An applicant must assess the number and type of customers he/she expects to sell to. This is done very well and we particularly like the different phases of the plan.

5	MARKETING
5·1	Primary schools · tuckshops · special events eg. sports days, fetes · gate sales during break time using trolley / cycle trailers.
5·2	Distribution means are three-fold:- · by trolley to nearby schools · by cycle trailers to further schools · by car to furthest point. The salesman will be responsible for distribution as well as creation of new markets, while the partners and labourers will also be required to distribute/ sell wherever a need arises.
5·3	Scale of production: During the first phase of the development of the business - (current), the scale of production was 1000 × 100 gm. packets per month. In phase II (proposed) the monthly production is at 4000 × 100 gm. packets per month with possibilities for increases.

There is good evidence of a lot of thought behind this section and this gives us confidence that the business is well researched and therefore more likely to succeed. The presentation of the information is clear and concise. As non-technical bankers we can understand it easily and again this gives us confidence in the business.

5.4	Fluctuation: It has been observed that the product will be marketed eight months per year due to schools breaking for holidays.
6	RESOURCES
6.1	Premises: Large room. To be demarcated into office, production area, storage room, pantry. Sinks to be installed.
6.2	Personnel: Manager - To manage/help with selling and distribution/ P.R. / Typist. Salesman - To promote/seek orders/ P.R./also process when free or when orders bulky. Book-keeper - To keep records/also help with promotion and P.R. To be sent for training in book-keeping. Labourers (3) - To process i.e. pop the corn, prepare syrup, dip corn in syrup, sprinkle with roasted peanut pieces. Also assist with distribution and selling whenever a need arises. On-the-spot training

Good thought to the market being present for only eight months per year. This is an important constraint which must be planned for. Again lots of detail which is clearly presented and shows that the applicant has thought seriously about the plan.

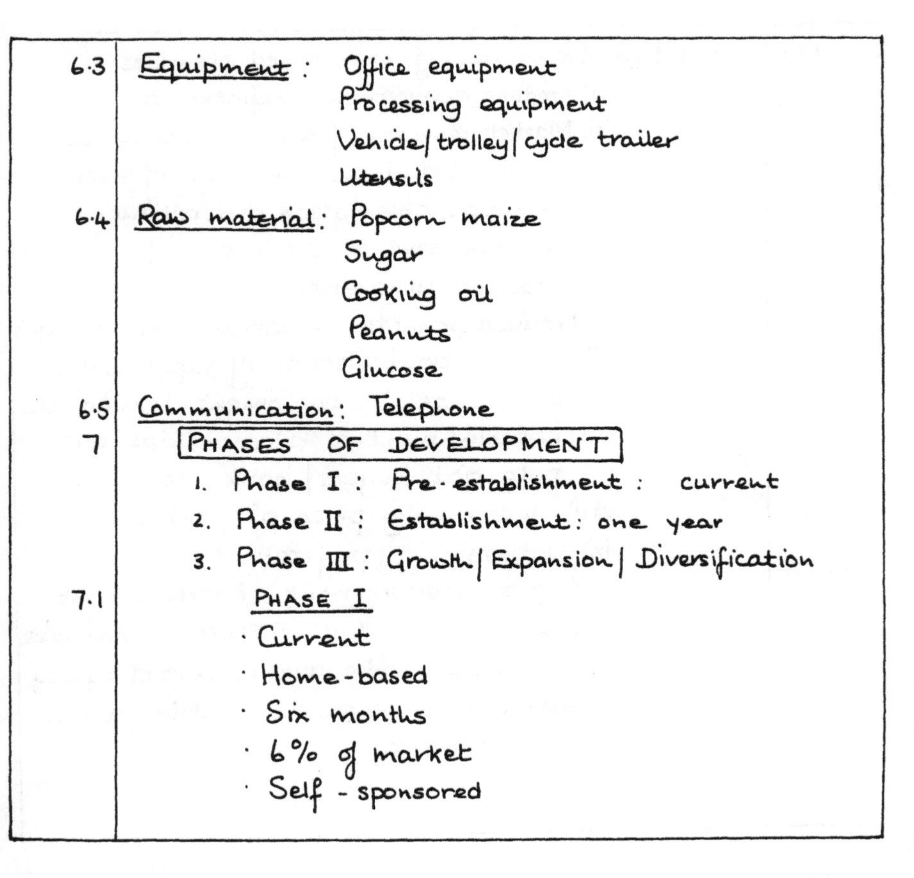

6.3	Equipment: Office equipment Processing equipment Vehicle/trolley/cycle trailer Utensils
6.4	Raw material: Popcorn maize Sugar Cooking oil Peanuts Glucose
6.5	Communication: Telephone
7	PHASES OF DEVELOPMENT 1. Phase I : Pre-establishment : current 2. Phase II : Establishment: one year 3. Phase III : Growth/Expansion/Diversification
7.1	PHASE I · Current · Home-based · Six months · 6% of market · Self-sponsored

Self sponsorship for phase 1 shows commitment to the business by the applicant and confidence in its success. This also makes us more confident about giving a loan.

7·1·1	Activities i) Renovation of home-based premises. ii) Product development / refinement. iii) Market research / feasibility study using varying techniques like trend of sales, comments, observation and attitude / reaction towards product iv) Product awareness v) Product promotion – samples / visits to schools vi) Identification / selection of suppliers of :- raw material, equipment, trailers etc vii) Establishment of relationships with potential buyers / bank etc. viii) Determining price of product. ix) Determining profitability x) Capital development. Exercise cost $ 4000-00. Got $ 5000·00 from sales. This amount to be used as part of equity contribution when negotiating a loan with the bank.

Good attention to the details needed in the planning phase. This information will be invaluable when the applicant approaches a bank or other lender for funding. A very professional approach.

7.2	PHASE II · Establishment · To operate from industrially designed premises. · One year · This business plan is for this phase · 15% of market to be covered i.e. to supply 15 schools – as per current arrangements/orders Activities · Renovation of newly acquired premises. To be rented. Larger room. · Registration of enterprise. · Purchase new/larger equipment. · Hire and train staff.
7.3	PHASE III · After the first year · Potential growth of enterprise covering more schools – 20%-25% · Possibility of purchasing better vehicle

Initial market share is not too optimistic. Plans for growth seem realistic and the delay in purchasing a new vehicle until the business is better established shows care and realism which, as lenders, we like.

Possible diversification should there be competition or new products to be developed. Possibility of supplying supermarkets.

8 PRODUCTION COSTS (Per day)

8·i FIXED COSTS

	$	c
Rent + Water: $100 ÷ 30 days	3	33
Labour: Allowances: $100 x 3 ÷ 30	10	00
Wages: $100 x 3 ÷ 30	10	00
Loan repayment: $1787 ÷ 365 days	4	80
Maintenance: 10% of equipment cost per year		48
Depreciation: 10% of equipment cost per year		48
SUB TOTAL	$ 29	09

ii) VARIABLE COSTS

a) Raw material

	$	c
Popcorn maize: 10 kg bag @ 50c kg	5	00
Sugar: 5 Kg @ $1·00 kg	5	00
Oil: 500 ml @ $2·50 per litre	1	25
Peanuts: 5 kg at 50c kg	2	50
Glucose: 500 ml for $2·00	2	00
SUB TOTAL	$ 15	75

Good to see plans for other products/customers in case the competition is stronger than expected. This is a very important consideration which is not always present in business plans. Costs appear to be accurate and are clearly presented.

It is important to include both wages and profits in business costing. Too often people undervalue their own time or pay themselves too little. It is important to remember that the profits belong to the business and are not the wages of the owner.

	$	c
b) Packaging		
Polypropylene: 100 metres ÷ 30 days ÷ 200 · 100gm pkts	0	01
Labels: 3000 ÷ 200 pkts ÷ $120		13
Ribbons: 50 metres x 25c ÷ 200 pkts		20
SUB TOTAL		34
c) Electricity		
$100 ÷ 30 days ÷ 24 hours x 6 hours		42
SUB TOTAL		·42
d) Transport		
$500 ÷ 30 days	16 ·	66
SUB TOTAL	16 ·	66

TOTAL OF PRODUCTION COSTS
= $29·09 + $15·75 + $0·34 + $0·42 + $16·66
= $62·26

+ 33⅓ % profit
= $62·26 + 20·75
= $83·01

Clearly presented, looks to be profitable. We would need to confirm the figures given for expenses at a later date.

N° of units produced per day	=	200
Price per unit	=	70 c
Income per day	=	200 pkts x 70c
	=	$ 140·00
Profit per day	=	$ 140·00 – 83·00
	=	$ 57·00
Profit per month	=	$ 57·00 x 30 = $1710
Profit per year	=	$ 1710 x 12
	=	$ 20520

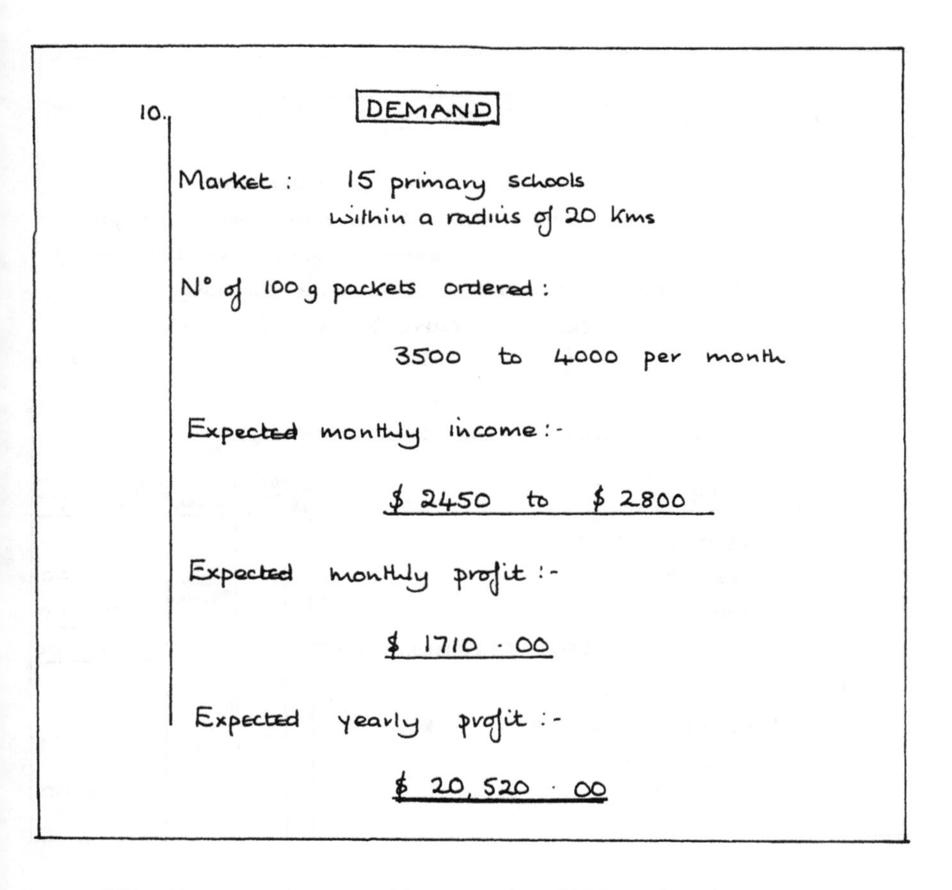

It would be better to show on this page what 4000 packets is as a percentage of the total expected demand

9 CAPITAL REQUIREMENTS

PHASE II

ESTABLISHMENT OF BUSINESS

ONE YEAR (Mainly applicable in 1st year amended for subsequent years)

N.B. Amounts presented in Zimbabwe currency

Rates of exchange currently are approximately:-

£1 = Z$ 7

U.S. $ = Z$ 2.5

Monthly and annual amounts shown

	ITEM	Qty No	Amount Price/Unit	TOTALS Monthly	TOTALS Annually
9.1	ESTABLISHMENT				
	Registration fees			ONE TIME PAYMENT	500.00
	Legal fees				500.00
	SUB TOTAL: ESTABLISHMENT				1000.00
9.2	FIXED ASSETTS				
9.2.1	Premises: Renovation/ Sinks/security bars etc.				1000.00

Good to see registration and legal fees included when they are needed. These are often forgotten.

9.2.2 Equipment				
i) Office				
Furniture	Varied			180 . 00
Typewriter (s/h)	1			75 . 00
Telephone installation				200 . 00
SUB TOTAL.				455 . 00
ii) Processing			ONE TIME PAYMENT	
Large table	2	$ 60		120 . 00
Large saucepans	3	$ 20		60 . 00
Large frying pans	2	$ 15		30 . 00
Hot plate	1	$ 100		100 . 00
SUB TOTAL				310 . 00
ITEM	QTY/No	AMOUNT PRICE/UNIT	MONTHLY	ANNUALLY
Sieve/ large	2	$ 10		20 . 00
Scale	1	$ 75		75 . 00
Spoons	6	$ 2		12 . 00
Perforated trays	5	$ 10		50 . 00
Pestle and mortar	1	$ 20		20 . 00
Heat sealer	1	$ 100		100 . 00
SUB TOTAL				277 . 00
SUB TOTAL				2042 . 00

iii) Transport/Vehicles/Distribution				
Car: To use family car – in this phase: Datsun S/W.				
Trolley: already acquired				
Cycle trailers:	3	$ 100		300 · 00
10% maintenance	3	$ 10		30 · 00
SUB TOTAL				330 · 00
SUB TOTAL: FIXED ASSETS.				2372 · 00
9·3 OPERATIONAL COSTS				
i) Salaries/Wages/Allowances				
Manager (To get allowances during the first year)	1	$ 100	100	1200 · 00
Salesman (To get allowances during the first year)	1	$ 100	100	1200 · 00
Book-keeper (To get allowances during the first year)	1	$ 100	100	1200 · 00
Labourer	3	$ 100	300	3600 · 00
SUB TOTAL.			600	7200 · 00

It would be useful to have a note saying how much product is to be distributed by car or bicycle trailer. Section 5.2 of the business plan tells us that the applicant needs both means of transport. Since the car is for longer journeys this might not be cost-effective and the business should then concentrate on nearby customers.

ITEM	QTY/NO	PRICE PER UNIT	TOTALS MONTHLY	TOTALS ANNUALLY
ii) Raw material				
Popcorn maize	300kg	50c kg	150.00	1800.00
Sugar	150kg	$1.00kg	150.00	1800.00
Vegetable oil	10 ltrs.	$2.00lt	20.00	240.00
Peanuts	150 kg	50c Kg	75.00	900.00
Glucose	10 ltrs	$4.00ltr	40.00	480.00
SUB TOTAL			$435.00	5220.00
iii) Packaging				
Polypropylene	100 m.	$1.00m	100.00	1200.00
Labels: gummed, decorated	3000	$40 per 1000	120.00	1440.00
Ribbons	50 m	25c	12.50	150.00
SUB TOTAL			232.50	2790.00
iv) Rent and Water			100.00	1200.00
SUB TOTAL			100.00	1200.00
v) Electricity			50.00	600.00
SUB TOTAL			$50.00	$600.00

The technical evaluators question the time needed to tie ribbons on each packet of product. This might slow down the process excessively so that the expected amount of product could not be made each day.
A process description, recipe and quality control details would be helpful to assess these figures properly. The equipment appears to be correct for this scale of operation and the materials listed here are adequate.

vi) Transport : car:.				
Fuel and repairs	1000kms	50c	500.00	6000.00
Trolleys \ cycle trailers				
10% maintence				
amount on 9.2				
SUB TOTAL.			$500.00	$6000.00
vii) Telephone			100.00	1200.00
SUB TOTAL			$100.00	$1200.00
viii) Stationery: lump sum			100.00	1200.00
SUB TOTAL			$100.00	$1200.00
ix) Postage			50.00	600.00
SUB TOTAL			$50.00	$600.00
SUB TOTAL OPERATIONAL COSTS			$2167.50	$26010.00

TOTAL OF BUDGET/CAPITAL REQUIREMENTS

	Monthly	Annually
Establishment	–	$ 1000-00
Fixed assets	–	$ 2372-00
Operational costs	$ 2167-50	$ 26010-00
Total in first year		$ 29382·00

Capital needed = $ 29382·00

Equity contribution = $ 14382-00

Loan required from bank = $ 15000-00

Period of repayment = 36 months

The equity contribution from the applicant comes from the first profits. This shows that she is willing to take the initial risks and this makes us more confident to lend her the funds that she requests. The repayment period is reasonable.

The cashflow forecasts on the following pages show clear evidence of a lot of thought and careful planning for each phase of the business. We also note the variable sales each month as demand from school children changes. This is well done and well presented.

1st YEAR	CASH-FLOW FORECAST : JAN – JUNE 1991						
RECEIPTS	JAN	FEB	MARCH	APRIL	MAY	JUNE	TOTALS
Cash sales	200	600	600	3000	400	800	5600
Cash from debtors	800	1400	2000	700	400	2000	7300
Capital	20000				1000		21000
Total receipts	21000	2000	2600	3700	1800	2800	33900
PAYMENTS							
Payments to Creditors	2500	6000	400	300	500	300	10000
Salaries, wages, allowances	600	600	700	600	600	600	3700
Rent / Water	100	100	100	100	100	100	600
Repairs			20			10	30
Electricity	50	60	50	20	40	60	280
Postage	30	20	25	36	10	25	146
Stationery	100	80	50	70	30	80	410
Transport	300	400	500	200	300	400	2100
Telephone	50	60	100	120	60	70	460

	JAN	FEB	MAR	APR	MAY	JUNE	TOTALS
Professional fees	500 (Registration	500 Lawyer					1000
Capital payments	Period of grace →			416	416	416	1248
Interest charges	Period of grace →			62	62	62	186
Other		Training 450					450
TOTAL PAYMENTS	4230	8270	1945	1924	2118	2123	20610
NET CASHFLOW	16770	−6270	655	1776	−318	677	13240

1st Year	CASH	FLOW FORCAST	JULY – DECEMBER 1991				
RECEIPTS	JULY	AUG	SEPT	OCT	NOV	DEC	
Cash sales	300	1500 (25% rise)	300	400	1560 } 100 % rise	200	4260
Cash from debtors	1500	2000	2000	1000	4000 }	600	11100
Capital Introduced	5000	3000					8000
TOTAL RECEIPTS	6800	6500	2300	1400	5560	800	23360
PAYMENTS							
Payments to Creditors	1000	600	900	500	1200	500	4700
Salaries, Wages Allowances	600	700 (over time)	600	600	800 (over time)	600	3900
Rent/Water	100	100	100	100	100	100	600
Repairs		800 (Car)			30		830
Electricity	50	100	75	60	110	40	435
Postage	30	80	75	40	90	10	325
Stationery	40	70	60	40	80	25	315
Transport	300	600	400	400	600	100	2400
Telephone	60	75	80	80	100	50	445
Professional Fees							
Capital payments	416	416	416	416	416	416	2496

Interest charges	62	62	62	62	62	62	372
Other	–	–	–	–	—	–	–
Total payments	2658	3603	2768	2298	3588	1903	16818
Net Cash Flow	4142	2897	-468	-898	1972	-1103	6544
1st Year Jan – June ANNUAL CASH FLOW	16770	-6270	605	1776	-318	677	13240
	4142	2897	-468	-898	1972	-1103	6542
	20912	-3373	137	878	1654	-426	19782

2nd Year CASH FLOW FORECAST : 2nd YEAR
- Increase of 75% due to demand
- Supplying 21 schools i.e. 21%
- Also supplying two nearby supermarkets

January – June 1992

RECEIPTS	JAN	FEB	MAR	APR	MAY	JUNE	TOTALS
Cash Sales	500	700	400	2000	500	200	4300
Cash from debtors	2000	3500	4000	2500	3600	4500	20100
Capital							
Total receipts	2500	4200	4400	4500	4100	4700	24400
PAYMENTS							
Payments to Creditors	800	600	700	600	400	900	4000
Wages/Salaries *	1550	1550	1550	1550	1550	1550	9300
Rent/Water	105	105	105	105	105	105	630
Repairs				CAR 2000		30	2030
Electricity	50	60	70	40	30	60	310
Postage	20	45	80	60	40	30	275
Stationery	80	90	40	30	50	60	350

Transport	600	700	500	500	300	500	3100
Telephone	50	70	80	90	60	50	400
Professional fees		1000					1000
Capital payments	416	416	416	416	416	416	2496
Interest charges	62	62	62	62	62	62	372
Other							
TOTAL PAYMENTS	3733	4698	3603	5453	3013	3763	24263
NET CASH FLOW	-1233	-498	797	-953	1087	937	137

* Sharp rise in salaries - manager - $ 350 p.m.
- salesman - $ 300 p.m.
- book keeper - $ 300 p.m.
- labourers - $ 200 each p.m.

Rent/ Water also increased by 5%

Had 2 set-backs - sued by one of the suppliers
- car needed major repairs.

2nd YEAR	CASH FLOW FORECAST					
RECEIPTS	JULY	AUG	SEPT	OCT	NOV	DEC	TOTALS
Cash from sales	400	300	400	600	800 *	200	2700
Cash from debtors	4000	3000	4500	4000	6000	2000	23500
Capital							
Total receipts	4400	3300	4900	4600	6800	2200	26200
PAYMENTS							
Payment to creditors	700	900	600	400	1000	600	4200
Wages/salaries	1550	1550	1550	1550	over time 2000	1550	9750
Rent/Water	105	105	105	105	105	105	630
Repairs			50				50
Electricity	70	50	40	60	100	30	350
Postage	40	20	30	40	60	50	240
Stationery	40	50	50	70	80	20	310
Transport	600	700	500	500	800	200	3300
Telephone	60	60	60	50	70	70	370
Professional fees							
Capital payments	416	416	416	416	416	416	2496
Interest charges	62	62	62	62	62	62	372

Other							
Total Payments	3643	3913	3463	3253	4693	3103	22068
Net Cash Flow	757	−613	1416	1347	2107	−903	4111
2nd YEAR Annual Cash Flow	−1233	−498	797	−953	1087	937	137
	757	−613	1416	1347	2107	−903	4111
	−476	−1111	2213	394	3194	34	4248
*	Increased sales by almost 25%						

Business plan 3: Mixed fruit chutney

Overall assessment
A clearly presented, well thought-out introduction. Good level of detail and easy to identify with the proposal.

Mixed Fruit Chutney
Production as a small business in
Nepal

Nepalese' normal meals include the rice, dal (boiled beans), vegetables cooked into curry and some pickles or Chutney. In Nepalese language it is called as Dal, Bhat, Tarkari and Achar. Achar (Chutney) is of two types, sweet and sour with hot spices. The word Chutney is thought to come from India thus used mostly in the southern part of Nepal where as in the mid-hills the term Achar is used.

Achar is made of fruits with spices and sugar to give the desired taste. The diced and cut pieces of fruit as well as vegetable are mixed with spices such as salt, coriander seeds' powder, cumin seeds' powder, ginger, garlic, chille powder and roasted fennel seeds, mustard seeds, etc. To make the final product look like a paste texture, the some of the fruits and onion are cooked by simply boiling and mix with spices after draining the water. Then mix starch dissolved in vinegar, sugar or molasses and fruit pulp with some acid. Cook for some time and packed in sterilised bottles as well as in plastic cups. The packaging is done while the Chutney is still hot to prevent the mould growth after packing. (Annex - i). Different fruits are used in different combination for different taste and also due to the availability of fruits due to different seasons. One combination is Apple and Pear or Apricot and orange mainly during winter season while in summer another combination is made of mango, pineapple and plum. (Annex - ii).

The technology behind this Chutney production is not complicated and most of the equipments required are simple, such as, stainless steel cooking pots, bowls, manual blender for making the fruit pulp.

Regarding the raw material, fruits are the major raw material which are found in the market. Most of the fruits are transported in the city because of excess in the local area where they are grown, as the local people can not consume all and also because of the income for those who grow. The vinegar and Acetic Acid are another things which are mostly imported products from neighbouring countries, vinegar comes from China mainly for the Chinese Restaurants and Acetic Acid from India as most of the things.

This Achar or Chutney is taken during meals to create appetite and spicy sweeten the mouth for taking the food. Another application of this Chutney is, it very well go with the snacks like samosa, deep fried or griddle cooked rotis, taken during afternoon tea break or after office/college on the way to home and some take at home as take away. But the chutney is a problem for take away while bùying samosa of rotis as a take away because of its semi liquid nature and to provide in small

quantities by the shop keepers, so many people will just buy the dry or fried products leaving the chutney behind. The snacks will be nice to eat if they had chutney at home, even if they have to buy the snacks. This Chutney is also served with the crackers, in Restaurants.

The Chutney can also be used as bread spread, for those who do not like to stick with the same taste of jam and butter.

The final product Chutney has the shelf life of up to six months after packing. There are two ways of extending theshelf life of the product, by boiling or cooking and packing when it is hot, and another way without cooking but treating or adding Potassium Sorbate. For making it simple, the product is cooked and packed in this production. Cooking also carried out in the improved institutional stove made with mud and brick for using firewood. Electricity or gas is expensive for this type of small business till now.

LOCATION :

The production centre will be located at a near by town very accessible to the city or urban area, Kathmandu, the capital of Nepal.
Possibilities - Near Balaju or Thankot, etc.

TARGET PEOPLE :

The market for this product is targeted to the urban people, mostly the working women who do not have time to prepare chutneys at home, and some of the housewives.

There are many people who come from outside the capital who usually take meals at restaurants, so restaurants are another market for the sale of this product.

Production Strategy :

Based on the market share estimation , it was estimated that the Chutney has to be produced at the rate of 35 kg per day, assuming to cover 2 % of the households (total no. of household 100,000)and 10 % of the existing restaurants in the city, that is 20 restaurants,(Annex - iii).

Two out of 20 restaurants will be approached at the first year by direct selling and for the households, sales through the existing retail shops, the supermarket.

Personnel :

The business supposed to hire two persons on wages and one manager or the owner, i.e., three persons are working. The salary for the manager is Rs.1500 per month and for the hired persons is Rs.40 per day per person.

Cost of Production :

Normally the Chutneys or pickles are sold in bottles, but here about half of the production will be in bottles and half will be packed in plastic cups with foil cover. Bottles are of 500 gm. size while plastic cups are of 200 gm.

Assuming that there are 220 working days per year, the cost of a bottle is Rs.23.24 and the cost of plastic cup is Rs.8.39.

The selling price is determined as Rs.27.50 for a bottle and Rs.10.00 for a cup. The existing prices for such bottle is Rs.25 to Rs.30 with only 400 gm. The plastic cup is a new packaging material for chutneys which provide opportunity for people with lower income which extends the market (Annex -iv).

This product is however a new product in market. But it has to compete with the existing practice and some of products which are different. Some sour and one fruit pickles are sometimes found in the market which come from India mostly.

People have the practice of making their own chutney or Achar, is another type of competition. Also the restaurants are making their own product for serving people. They make small quantities time and again so they might prefer to buy instead. The restaurants will be approached and allowed to taste at the beginning.

We like the diversification of potential markets and the realistic levels of initial demand. However, in view of other products already being made, we feel that greater attention needs to be given to promotion of this product and the costs associated with this.

The speciality of this product is it gives different tastes of different fruits, its wider application apart from taking with the meals, and because it is sweet children like it most. Now more and more women in the city are busy with their work besides household works so they do not have time to prepare by themselves.

Production months :

It is always good to produce a new product just before the festival which lies some time in September. So it is aimed to start producing in the month of August, but only 50 % of total production for the month will be produced when there are less availability of raw materials. Due to the availability of raw material, the fruits Jan, Feb and Jul also have 50 % production while Mar there will not be any production. The rest of the months will have full production but in Dec only two third of production. The raw material supply for the months Jan, Feb and Jul will be bought before-hand and dried for the use later.

There will be five working days for production and one day per week is spent in the selling. The products will be delivered by hiring a three wheeler. Due to the new products, it will be delivered to the shops also but normally the payment will be about a month later, so it has to be on credit.

Investment and Loan from Bank :

It was estimated that there will not be any income for upto two months at the beginning. The start up expenses for the two months is calculated as Rs.72817.71 (Annex - v).

It was intended to take loan of Rs.50,000.00 from the bank and the remaining amount of Rs.22817.17 will be owner's investment. Bank provide three months grace period for the repayments of interest as well as the capital repayment. It was aimed to repay the capital as well as interest within one year after the grace period. The current interest rate of the bank money is 15 percent.

Weaknesses :

The production of Chutney in the urban ares of Nepal has some strengths as it can get the market, availability of fruits while there are some weaknesses such as it has to depend more on the importation of materials like bottles, plastic cups and the vinegar, acid. Sometimes, it will have problem on fruit availability as in the case of mango, some year there will plenty while some years the mango production on trees will be less.

Business status :

The cashflow and profit charts show that it is profitable. It was aimed to reach 5 percent of total households and 15 percent of the restaurants within the next two years.

Promotion :

First year concentrate by selling without promotion but for increasing the market share some activities for promotion will be carried put. About 2 percent of total profit will be spent for advertising and good attractive label.

Packaging material recovery :

It was was hoped to use some of the bottles used in the first year in coming years. But for the first year it will be imported/bought from dealers in bulks. It was hoped to recover about 5 percent of the used bottles through the bottle collectors.

We do not agree that promotion should take place after the first year. The product should be heavily promoted prior to/during its introduction to generate the initial sales. The weakness in material supply could be a significant problem. A range of products to take account of raw material fluctuations could be a possible solution.

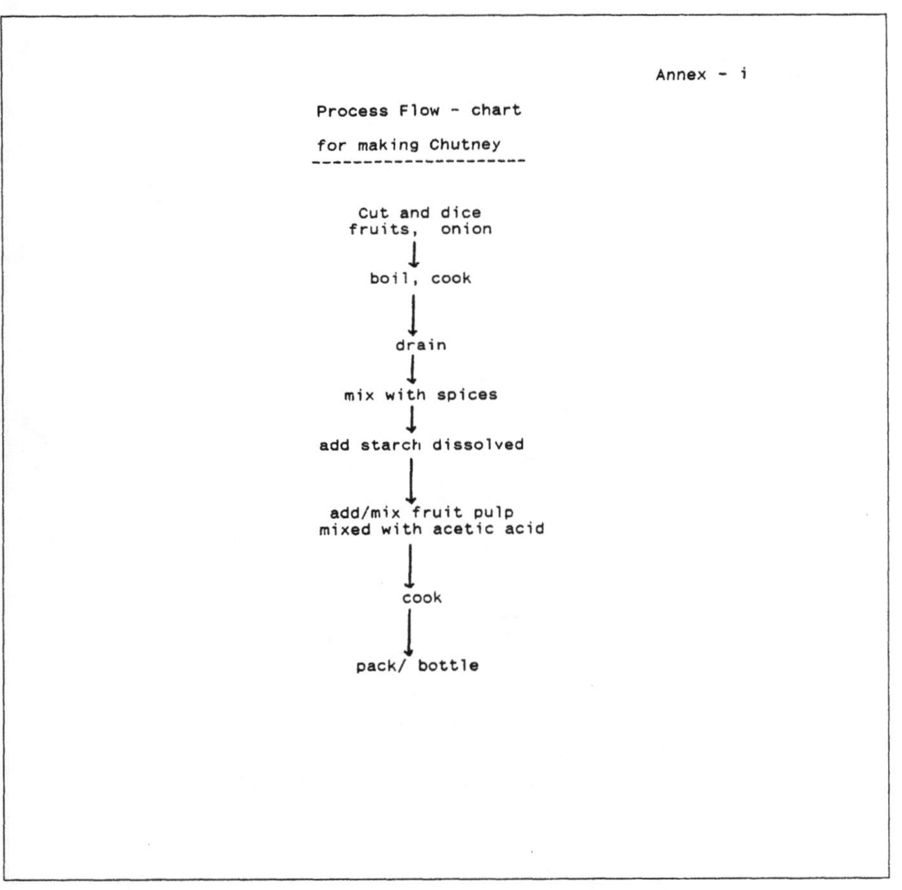

Annex - i

Process Flow - chart

for making Chutney

Cut and dice fruits, onion

↓

boil, cook

↓

drain

↓

mix with spices

↓

add starch dissolved

↓

add/mix fruit pulp mixed with acetic acid

↓

cook

↓

pack/ bottle

A useful diagram which would benefit from the inclusion of quality control procedures.

Annex - ii

Chart showing the availability of fruits in different months

Name of fruit	Mar	Apr	May	Jun	Jul	Aug	Sep	Oct	Nov	Dec	Jan	Feb
Apricot/Peach						----	----	----	----	----	----	
Pear						----	----	----	----	----		
Apple							----	----	----	----	----	
Mango		----	----	----								
Pineapple		----	----	----								
Plum			----	----	----							
Orange								----	----	----	----	

Annex - iii

Market Share (Demand Calculation)

Total number of households in the city area 100,000.00
Assuming 50 % of household buy Chutney 50,000.00

My target is to reach 2 % of the households - 1,000.00
Suppose 1 household buy 1 bottle in 2 weeks,
or 2 bottles per household per month 2000.00
bottles of 500 gm.
i.e. 1000 kg.
Direct selling to Restaurants,
Number of Restaurants - 20
Estimated to reach 10 %
of the restaurants, i.e. 2
Suppose 1 Restaurant serving 50 customers for meals in a day
and each customer served 25 gm of Chutney
that comes to 1.25 kg per restaurant per day.
Total quantity required for two restaurants will be
2 x 1.25 = 2.5 kg per day
restaurants open 30 days per month which requires
2.5 x 30 = 75 kg or 150 bottles per month

The total estimated demand
1000 kg for hh + 75 kg for rests. = 1075 kg.
i.e., demand for one day is 1075/30 = 35.8 kg.
35 kg.

Production Strategy

Month	Production
August	50 %
September	100 %
October	100 %
November	100 %
December	75 %
January	50 %
February	50 %
March	-
April	100 %
May	100 %
June	100 %
July	50 %

The production strategy and seasonality chart on the previous page show that careful attention had been given to some very real problems of raw material supply in this proposal. The applicant could also consider storage of part-processed fruits in drums to even out production rate throughout the year.

Costs of Equipment

	Unit Cost	Amt.Rs.	
Cooking pot Stainless Steel	700.00	1400.00	
Cutting Boards (Wooden)	20.00	40.00	
Mixing Bowls Stainless Steel	200.00	400.00	
Knives	25.00	50.00	
Weighing Scale	1500.00	1500.00	
Spatula (Wooden)	15.00	15.00	
Bucket Plastic	50.00	100.00	
Blender (Indian model)	800.00	1600.00	
Barrels	200.00	400.00	
Total cost			5505.00

Cost Raw Materials
35 Kg. of production per day

Materials	Quantity kg/li	Rs./Unit	Amt.Rs.	
Apricot/Peach	9.000	15.00	135.00	
Pear	7.000	5.00	35.00	
Onion	3.000	17.00	51.00	
Salt	0.300	6.00	1.80	
Ginger	0.100	10.00	1.00	
Spices	0.500	25.00	12.50	
Starch	0.210	15.00	3.15	
Vinegar	1.150	15.00	17.25	
Sugar	8.850	15.00	132.75	
Apple (puree)	5.800	15.00	87.00	
Glacial Acetic Acid	0.455	500.00	227.50	
Total cost				703.95 704.00

The cost calculations and cashflows given on the following pages are very well prepared and show that a great deal of thought has been given to this proposal. This makes us confident that with some changes, it may become a viable, sustainable business.

Cost of Production

It is assumed that there are 220 Working days per year

Fixed Costs

1. Rent	12000.00	
2. Salary	18000.00	
3. Administration & Registration	6000.00	
4. Interest of Loan	7500.00	
5. Depreciation (50 % of Equip)	2752.50	
Total		46252.50

Variable Costs

6. Wages	17600.00	
7. Raw Materials	154880.00	
8. Fuel (Firewood)	4400.00	
9. Transport	7000.00	
10. Maintenance & Repair	550.50	
Total		184430.50
Total Production Cost For Year		230683.00

Total Kilogram of Production	7700.00	
Total Production Cost per one Kilogram (without packaging)	29.96	
Unit cost with Packaging		
Bottles of 500 gm. capacity		
Product cost	14.98	
Bottle + Cap & label	8.50	
Total	23.48	
Selling Price per Bottle		Rs.27.50
Plastic Cups 200 gm capacity		
Product cost	5.99	
Cup + lid + label Rs	2.50	
Total	8.49	
Selling Price per Cup		Rs.10.00

Start up Expenses (for two Months)

Fixed Costs

1. Equipments		5505.00	
2. Rent @Rs.1000.00/month		2000.00	
3. Salary @Rs.1500.00/month		3000.00	
4. Administration		500.00	
5. Registration fee		2000.00	
			13005.00

Variable Costs

6. Raw Materials @Rs.704.00/day x25days/month		35200.00		
7. Packaging				
Bottles 876 X 8.50	7446.00			
Cups 1084 x 2.50	2710.00			
		10156.00		
8. Transportation @Rs.200/day x 4 days/month		1600.00		
9. Fuel @Rs.20/day x 25days/month		1000.00		
10. Labour @Rs.40/day x 2persons x 25days/month		4000.00		
11. Maintenance & Repair (10% of Equip.cost)		55.05		
Total Expenses			52011.05	
				65016.05
Contingencies	5 % of F C			2600.55
	10 % of V C			5201.11
		Totals		72817.71

Month JAN Budget Rs	£	Month FEB Budget Rs.	£	Month MAR Budget Rs	£	Month APR Budget Rs.	£	Month MAY Budget Rs.	£	Month JUN Budget Rs.	£	Month JUL Budget Rs	£	Totals Budget Rs.	£
5746.25	78.72	5746.25	78.72	0.00	0.00	11492.50	157.43	11492.50	157.43	11492.50	157.43	5746.25	78.72	100559.38	1377.53
25858.13	354.22	0.00	0.00	17238.75	236.15	0.00	0.00	34477.50	472.29	34477.50	472.29	34477.50	472.29	267200.63	3660.28
0.00	0.00	0.00	0.00	0.00	0.00	0.00	0.00	0.00	0.00	0.00	0.00	0.00	0.00	72817.71	997.50
31604.38	432.94	5746.25	78.72	17238.75	236.15	11492.50	157.43	45970.00	629.73	45970.00	629.73	40223.75	551.01	440577.72	6035.31
0.00	0.00	0.00	0.00	0.00	0.00	17700.57	242.47	17700.57	242.47	26550.85	363.71	0.00	0.00	172580.53	2364.12
8500.00	116.44	0.00	0.00	0.00	0.00	8500.00	116.44	0.00	0.00	17750.00	243.15	0.00	0.00	88750.00	1215.75
251.43	3.44	251.43	3.44	502.85	6.89	502.85	6.89	502.85	6.89	502.85	6.89	251.43	3.44	4902.80	67.16
2500.00	34.25	2500.00	34.25	1500.00	20.55	3500.00	47.95	3500.00	47.95	3500.00	47.95	2500.00	34.25	35500.00	486.30
1000.00	13.70	1000.00	13.70	1000.00	13.70	1000.00	13.70	1000.00	13.70	1000.00	13.70	1000.00	13.70	12000.00	164.38
500.00	6.85	500.00	6.85	500.00	6.85	500.00	6.85	500.00	6.85	500.00	6.85	500.00	6.85	6000.00	82.19
400.00	5.48	400.00	5.48	0.00	0.00	800.00	10.96	800.00	10.96	800.00	10.96	400.00	5.48	7000.00	95.89
31.46	0.43	31.46	0.43	0.00	0.00	62.91	0.86	62.91	0.86	62.91	0.86	31.46	0.43	550.48	7.54
0.00	0.00	0.00	0.00	0.00	0.00	0.00	0.00	0.00	0.00	0.00	0.00	0.00	0.00	5505.00	75.41
781.25	10.70	781.25	10.70	781.25	10.70	781.25	10.70	781.25	10.70	781.25	10.70	781.25	10.70	7031.25	96.32
4166.67	57.08	4611.67	63.17	4611.67	63.17	4611.67	63.17	4611.67	63.17	4611.67	63.17	4611.67	63.17	40170.03	550.27
18130.81	248.37	10075.81	138.02	8895.77	121.86	37959.25	519.99	29459.25	403.55	56059.53	767.94	10075.81	138.02	379990.09	5205.34
13473.57	184.57	-4329.56	-59.31	8342.98	114.29	-26466.75	-362.56	16510.75	226.17	-10089.53	-138.21	30147.94	412.99	60587.63	829.97
32998.23	452.03	46471.80	636.60	42142.24	577.29	50485.22	691.58	24018.47	329.02	40529.22	555.19	30439.69	416.98	0.00	0.00
46471.80	636.60	42142.24	577.29	50485.22	691.58	24018.47	329.02	40529.22	555.19	30439.69	416.98	60587.63	829.97	60587.63	829.97

Cashflow forecast

CASHFLOW FORECAST FOR : August 1992 July 1992

RECEIPTS	Month AUG Budget Rs.	£	Month SEPT. Budget Rs.	£	Month OCT Budget Rs.	£	Month NOV Budget Rs.	£	Month DEC Budget Rs.	£
Cash Sales	5746.25	78.72	11492.50	157.43	11492.50	157.43	11492.50	157.43	8619.38	118.07
Cash from Debtors	0.00	0.00	17238.75	236.15	34477.50	472.29	34477.50	472.29	34477.50	472.29
Capital Introduced	72817.71	997.50	0.00	0.00	0.00	0.00	0.00	0.00	0.00	0.00
TOTAL RECEIPTS (a)	78563.96	1076.22	28731.25	393.58	45970.00	629.73	45970.00	629.73	43096.88	590.37
PAYMENTS										
Raw Materials	8850.26	121.24	17700.57	242.47	35401.14	484.95	35401.14	484.95	13275.43	181.86
Packaging Material	13500.00	184.93	13500.00	184.93	13500.00	184.93	13500.00	184.93	0.00	0.00
Fuel	251.43	3.44	502.85	6.89	502.85	6.89	502.85	6.89	377.13	5.17
Salaries/Wages	2500.00	34.25	3500.00	47.95	3500.00	47.95	3500.00	47.95	3000.00	41.10
Rent	1000.00	13.70	1000.00	13.70	1000.00	13.70	1000.00	13.70	1000.00	13.70
Administration	500.00	6.85	500.00	6.85	500.00	6.85	500.00	6.85	500.00	6.85
Transport	400.00	5.48	800.00	10.96	800.00	10.96	800.00	10.96	600.00	8.22
Maint. & Repairs	31.46	0.43	62.91	0.86	62.91	0.86	62.91	0.86	47.18	0.65
Capital Payments	5505.00	75.41	0.00	0.00	0.00	0.00	0.00	0.00	0.00	0.00
Interest Charges	0.00	0.00	0.00	0.00	0.00	0.00	781.25	10.70	781.25	10.70
Loan Repayments	0.00	0.00	0.00	0.00	0.00	0.00	4166.67	57.08	4166.67	57.08
TOTAL PAYMENTS (b)	32538.15	445.73	37566.33	514.61	55266.90	757.08	60214.82	824.86	23747.66	325.31
NET CASHFLOW (a-b)	46025.81	630.49	-8835.08	-121.03	-9296.90	-127.35	-14244.82	-195.13	19349.22	265.06
Opening Bank Balance	0.00	0.00	46025.81	630.49	37190.73	509.46	27893.83	382.11	13649.01	186.97
Closing Bank Balance	46025.81	630.49	37190.73	509.46	27893.83	382.11	13649.01	186.97	32998.23	452.03

Month JAN Budget	£	Month FEB Budget	£	Month MAR Budget	£	Month APR Budget	3	Month MAY Budget	£	Month JUN Budget	£	Month JUL Budget	£	Totals Budget	£
22985.00	314.86	22985.00	314.86	0.00	0.00	45970.00	629.73	45970.00	629.73	45970.00	629.73	22985.00	314.86	402237.51	5510.10
8850.29	121.24	8850.29	121.24	0.00	0.00	17700.57	242.47	17700.57	242.47	17700.57	242.47	8850.29	121.24	154880.00	2121.64
1000.00	13.70	1000.00	13.70	0.00	0.00	2000.00	27.40	2000.00	27.40	2000.00	27.40	1000.00	13.70	17500.00	239.73
13134.71	179.93	13134.71	179.93	0.00	0.00	26269.43	359.86	26269.43	359.86	26269.43	359.86	13134.71	179.93	229857.51	3148.73
57.14	0.78	57.14	0.78	0.00	0.00	57.14	0.78	57.14	0.78	57.14	0.78	57.14	0.78	57.14	0.78
1500.00	20.55	1500.00	20.55	1500.00	20.55	1500.00	20.55	1500.00	20.55	1500.00	20.55	1500.00	20.55	18000.00	246.58
1000.00	13.70	1000.00	13.70	1000.00	13.70	1000.00	13.70	1000.00	13.70	1000.00	13.70	1000.00	13.70	12000.00	164.38
251.43	3.44	251.43	3.44	0.00	0.00	502.85	6.89	502.85	6.89	502.85	6.89	251.43	3.44	4399.96	60.27
31.46	0.43	31.46	0.43	0.00	0.00	62.91	0.86	62.91	0.86	62.91	0.86	31.46	0.43	550.48	7.54
400.00	5.48	400.00	5.48	0.00	0.00	800.00	10.96	800.00	10.96	800.00	10.96	400.00	5.48	7000.00	95.89
500.00	6.85	500.00	6.85	500.00	6.85	500.00	6.85	500.00	6.85	500.00	6.85	500.00	6.85	6000.00	82.19
2083.33	28.54	2083.33	28.54	2083.33	28.54	2083.33	28.54	2083.33	28.54	2083.33	28.54	2083.33	28.54	18749.97	256.85
5766.22	78.99	5766.22	78.99	5083.33	69.63	6449.09	88.34	6449.09	88.34	6449.09	88.34	5766.22	78.99	66700.41	913.70
7368.49	100.94	7368.49	100.94	-5083.33	-69.63	19820.34	271.51	19820.34	271.51	19820.34	271.51	7368.49	100.94	163157.10	2235.03
157.29	2.15	157.29	2.15	0.00	0.00	314.57	4.31	314.57	4.31	314.57	4.31	157.29	2.15	2751.76	37.70
7211.20	98.78	7211.20	98.78	-5083.33	-69.63	19505.77	267.20	19505.77	267.20	19505.77	267.20	7211.20	98.78	160405.34	2197.33

Profit budget

PROFIT BUDGET FOR August 1992 To July 1993

	Month AUG Budget	£	Month SEPT Budget	£	Month OCT Budget	£	Month NOV Budget	£	Month DEC Budget	£
Sales (a)	22985.00	314.86	45970.00	629.73	45970.00	629.73	45970.00	629.73	34477.51	472.29
Less:Direct Costs										
Cost of Materials	8850.29	121.24	17700.57	242.47	17700.57	242.47	17700.57	242.47	13275.43	181.86
Wages	1000.00	13.70	2000.00	27.40	2000.00	27.40	2000.00	27.40	1500.00	20.55
GROSS PROFIT (b)	13134.71	179.93	26269.43	359.86	26269.43	359.86	26269.43	359.86	19702.08	269.89
Gross Profit Margin (b/ax100)	57.14	0.78	57.14	0.78	57.14	0.78	57.14	0.78	57.14	0.78
Overheads										
Salaries	1500.00	20.55	1500.00	20.55	1500.00	20.55	1500.00	20.55	1500.00	20.55
Rent	1000.00	13.70	1000.00	13.70	1000.00	13.70	1000.00	13.70	1000.00	13.70
Fuel	251.43	3.44	502.85	6.89	502.85	6.89	502.85	6.89	377.14	5.17
Maint. & Repairs	31.46	0.43	62.91	0.86	62.91	0.86	62.91	0.86	47.18	0.65
Transport	400.00	5.48	800.00	10.96	800.00	10.96	800.00	10.96	600.00	8.22
Administration	500.00	6.85	500.00	6.85	500.00	6.85	500.00	6.85	500.00	6.85
Interest Charges	0.00	0.00	0.00	0.00	0.00	0.00	2083.33	28.54	2083.33	28.54
TOTAL OVERHEADS(c)	3682.89	50.45	4365.76	59.80	4365.76	59.80	6449.09	88.34	6107.65	83.67
TRADING PROFIT (b)-(c)	9451.82	129.48	21903.67	300.05	21903.67	300.05	19820.34	271.51	13594.43	186.23
Less:Depreciation	157.29	2.15	314.37	4.31	314.37	4.31	314.37	4.31	235.78	3.23
NET PROFIT	9294.53	127.32	21589.30	295.74	21589.30	295.74	19505.97	267.21	13358.65	183.00

CASHFLOW FORECAST FOR	1992/93		1993/94	
	Totals			
RECEIPTS	Budget Rs.	£	Budget Rs.	£
Cash Sales	100559.38	1377.53	162906.20	2231.59
Cash from Debtors	267200.63	3660.28	432865.02	5929.66
Capital Introduced	72817.71	997.50	0.00	0.00
TOTAL RECEIPTS (a)	440577.72	6035.31	595771.22	8161.25
PAYMENTS				
Raw Materials	172580.53	2364.12	279580.46	3829.87
Packaging Material	88750.00	1215.75	143775.00	1969.52
Fuel	4902.80	67.16	7942.54	108.80
Salaries/Wages	35500.00	486.30	50587.40	692.98
Rent	12000.00	164.38	14400.00	197.26
Administration	6000.00	82.19	7200.00	98.63
Transport	7000.00	95.89	11340.00	155.34
Maint. & Repairs	550.48	7.54	891.78	12.22
Capital Payments	5505.00	75.41	2000.00	27.40
Interest Charges	7031.25	96.32	2343.75	32.11
Loan Repayments	40170.03	550.27	12500.01	171.23
TOTAL PAYMENTS (b)	379990.09	5205.34	532560.94	7295.36
NET CASHFLOW (a-b)	60587.63	829.97	63210.28	865.89
Opening Bank Balance	0.00	0.00	60587.63	829.97
Closing Bank Balance	60587.63	829.97	123797.91	1695.86

Business aspects

Small Business in the Third World, M. Harper, 1984, IT Publications, 103–105 Southampton Row, London WC1B 4HH, UK (ISBN 0 471 90474 0)

Entrepreneurship for the Poor, M. Harper, 1984, IT Publications, 103–105 Southampton Row, London WC1B 4HH, UK (out of print)

Doing a Feasibility Study: training activities for starting or reviewing a small business, Suzanne Kindervater (Editor), 1987, OEF International, 1815 H Street NW, 11th Floor, Washington, DC 20006, USA (ISBN 0 912917 07 5)

Consultancy for Small Businesses: the concept, training the consultants, M. Harper, 1976, IT Publications(ISBN 0 903031 42 6)

Monitoring and Evaluating Small Business Projects: a step by step guide for private development organisations, S. Buzzard and E. Edgcomb (Editors), 1992, PACT, 777 United Nations Plaza, New York, NY 10017, USA (ISBN 0 942127 00 5)

Marketing Strategy: training activities for entrepreneurs, S. Kindervater and M. Range, 1986, OEF International, 1815 H Street NW, 11th Floor, Washington, DC 20006, USA (ISBN 0 912917 08 3)

Technical aspects

Small Scale Food Processing: a guide to appropriate equipment, P. Fellows and A. Hampton, 1992, IT Publications, 103–105 Southampton Row, London WC1B 4HH, UK (ISBN 1 85339 108 5)

Food Cycle Technology Source Books, series aimed at women, covering various aspects of food processing. IT Publications (address above)

Making Safe Food, P. Fellows and V. Hidellage, 1992, From Technical Enquiry Unit, ITDG, Myson House, Railway Terrace, Rugby, CV21 3HT, UK

Traditional Food, P. Fellows (Editor), 1996, IT Publications (address above) (ISBN 1 85339 228 6)

Improving Small-Scale Food Industries in Developing Countries, W. Edwardson and C.W. MacCormac (Editors), 1986, IDRC Publications, PO Box 8500, Ottawa, Canada K1G 3H9 (ISBN 0 88936 398 6)

Appropriate Food Packaging, P. Fellows and B. Axtell, 1993, TOOL Publications, Sarphatistraat 650, 1018 AV Amsterdam, The Netherlands (ISBN 90 70857 28 6)

Processing Tropical Crops, J.J. Asiedu, 1989, MacMillan Press Ltd, London, UK (ISBN 0 333 44857 X)

Traditional and Non-traditional Foods, R. Ferrando, 1981, FAO Publications, Via delle Terme di Caracalla, Rome, Italy (ISBN 92 5 100167 7)

Food: the chemistry of its components, T.P. Coultate, 1984, The Royal Society of Chemistry, Burlington House, London W1V 0BN, UK (ISBN 0 85186 483 X).

Hygienic Design and Operation of Food Plant, R. Jowitt (Editor), 1980, Ellis Horwood Ltd, Cooper Street, Chichester PO19 1EB, UK (ISBN 0 85312 153 2) [NB for larger scale food processing]

Food and Drink – Good Manufacturing Practice: a guide to its responsible management, IFST, 5 Cambridge Court, 210 Shepherd's Bush Road, London W6 7NL, UK (ISBN 0 905367 08 1)

Food Poisoning and Food Hygiene, B. Hobbs and D. Roberts, 1987, Edward Arnold Ltd, 41 Bedford Square, London WC1B 3DQ, UK (ISBN 0 7131 4516 1)

How to score the checklist in Appendix 1

Score 5 points for YES answers and 0 points for NO answers to questions 2, 4, 7, 8, 10, 12, 14

Score 5 points for NO answers and 0 points for YES answers to questions 1, 3, 5, 6, 9, 11, 13.

The maximum score is 70 points and the higher you score the better are your chances of successfully setting up and operating a small food processing business.

www.ingramcontent.com/pod-product-compliance
Lightning Source LLC
Jackson TN
JSHW011411130125
77033JS00024B/964

* 9 7 8 1 8 5 3 3 9 3 2 3 5 *